Introduction to Clinical Ethics: Perspectives from a Physician Bioethicist

Saleem Toro

Introduction to Clinical Ethics: Perspectives from a Physician Bioethicist

Saleem Toro
Department of Cardiology
Houston Methodist DeBakey Heart
and Vascular Center
Houston, TX, USA

ISBN 978-3-031-30806-2   ISBN 978-3-031-30804-8 (eBook)
https://doi.org/10.1007/978-3-031-30804-8

This Springer imprint is published by the registered company Springer Nature Switzerland AG
The registered company address is: Gewerbestrasse 11, 6330 Cham, Switzerland

To my family, Edward, Aria, George, Fadi, and Carel.

Preface

When I was in medical school back in Syria, I vaguely remember hearing some of my professors talking about medicine being not only a science but also an art. Honestly, I did not quite get what those mentors meant by such a phrase. As a person who loved literature and played classical music for 14 years, I was somewhat aware of arts and aesthetics. To me, studying histology, microbiology, pathology, nephrology, cardiology, and the rest of medical school curriculum was close to being anything on earth except arts. Even a few years later, after starting our clinical rotations, I couldn't grasp that artistic dimension in taking a patient's history, performing a physical exam, and coming up with a differential diagnosis. I kept telling myself, what for heaven's sake are these hospital attendings talking about? Medicine is an art! The only thing I am doing these days is memorizing volumes and volumes of textbooks.

In 2011, the war started in Syria and dramatically changed our perception of life and the surrounding world. We learned to get accustomed to losing many things we loved, including friends and family members. We experienced suffering, pain, fear, despair, and death in his various ugly faces. Rounding on hospital wards in those days brought up everything except arts. There were a lot of tragedies and sadness. There were barely any lights or electric power. The shortage of water supply was an everyday experience, and besides the noise of the hospital's monitors, the only music around us was the bombardments, explosions, and the whizzes of bullets.

Despite that, my personal search for finding the art and wisdom in medicine continued throughout the following years. Though it passed through a period of quiescence, I would say, especially after starting my internship in internal medicine residency. The first two years of that journey were extremely demanding and left little time for reflecting on what I was doing. However, after some time, and after becoming more comfortable with the daily practice of medicine, most of the training physicians, including myself, started to ask deeper questions about the nature of what we do as healthcare professionals. And whether we are delivering good care to our patients. More profoundly, what is, in fact, the definition of a good quality care?

During my third year of internal medicine training, a physician-bioethicist gave us a wonderful lecture on health care ethics and the goals of medicine. As far as I can remember, this was my first encounter with what the art of medicine could have actually been. What is the philosophy of medicine, and how could someone study the humanistic aspect of medicine academically and methodically. However, I do also remember that many of my co-residents attending this lecture were neither, to be honest, attracted to the content, nor happy with me asking so many questions and prolonging the talk. Nevertheless, after finishing up the teaching session, I walked to the lecturer, and I spoke to him about how interesting these ideas are and the best way to start researching and reading more about the field of medical ethics.

Several years later, after formal studies in medical ethics, I present this introduction to my first work in this field. I hope to communicate some of the fascinating ideas I have learned during journey of medical research and patients care. Hence, this work aims to introduce physicians to the field of medical ethics, providing a brief reference on a broad spectrum of different subjects in the practice of medicine. I seek to explain the concepts to make them easier to understand. Perhaps, abstract thinking, which is the predominant model of thinking in philosophy, is one of the main reasons why many physicians find ethical arguments complicated, theoretical, and often unpractical. However, when professional ethicists write about medical ethics in a strict philosophical sense, their presentation of the ideas sometimes becomes too abstract and difficult to apply in day-to-day clinical practice, making the conversation between the two sides quite difficult. Therefore, I try to build some bridges between practicing physicians and medical ethicists, and I hope that the content will find resonance in the heart and minds of those interested in the field regardless of their background.

So, did I finally find the art in medicine? Perhaps the answer to this question is more complicated than yes or no. And to be quite honest, I am still learning to appreciate what medicine qua medicine is. I believe the difficult answer to this question lies in defining a clear philosophy of medicine. In essence, medicine is not purely a scientific field since the main goal of medicine is different from the main goal of science. Science aims to find practical truth, whereas medicine aims for patient healing. This healing is partially biomedical, achieved by applying scientific methods. However, our job as physicians is not confined to that aspect alone. The art of practice is present in the attempts to understand our patients as humans in the full human sense; through our unique relationship with them as healthcare professionals. This healing relationship is bounded by trust, fidelity, and vulnerability, within which we have to examine the phenomenology of illness and suffering. Knowing that we are also vulnerable and one day we might be experiencing that same suffering ourselves. This holistic approach is far from being solely scientific and requires the wisdom of understanding who we are as humans in a broader sense.

I seek to foster focus on teaching medical humanities in medical schools and residency programs. As a physician currently in fellowship training in cardiovascular medicine, I understand the importance of science and scientific progress in delivering care to our patients. The medical humanities can have a critical role in patient care, celebrating who we are as humans. When someone who is suffering asked for help, the healing relationship begins. I hope this account brings light and encouragement when engaging clinical ethical dilemmas.

Houston, USA   Saleem Toro

Acknowledgments

This book shares some of my thoughts and reflections on the ethical side of practicing medicine. Those reflections are not solely mine but rather the result of the collective experience and presence of many wonderful people who helped and supported me during the various stages of my professional career.

First and foremost, I would like to thank my mentor Dr. Gerard Magill, whose invaluable guidance, encouragement, and support throughout the writing process have been indispensable. Without him, this book could not have been possible. I am also sincerely grateful to Dr. Johan Bester for introducing me to the field and guiding me to pursue further studies in clinical ethics.

Outside the bioethics sphere, mentors and staff physicians whose practice and patient care formed the model of my future self have inspired me, enriched my career, and helped me become a better clinician. I want to acknowledge Dr. Riad Mortada, Dr. Martin Maron, Dr. Ethan Rowin, and Dr. Jaime Gerber.

I am also enormously grateful to Dr. Jennifer Ramsey for all her care and support during my residency training. I also owe a huge debt of gratitude to Dr. William Zoghbi for reviewing part of the manuscript and for his continuous mentorship and guidance during my cardiology fellowship training.

Furthermore, a special recognition extends to my wonderful friends, who have been a grace and blessing in my life. Without them, I could not have reached where I am at present. I would like to acknowledge Imad Jabbour, Dr. Tala Jarjour, Dr. Louis Saade, Dr. Nadeen Faza, and Dr. Serge Harb.

Many heartfelt thanks go to Dr. Hikmat Hannawi and his wife, Mouna Doudak, and to Joseph Hannawi and Abdul Hannawi for all their love and care.

Finally, I would like to express my deep appreciation to my small family, to Geroge Toro, who keeps me hoping; to Fadi Toro, who keeps me smiling; to Carel Toro, who keeps me dreaming. And to my parents, Edward and Aria, whose unwavering love and sacrifice have been the foundation of my life.

Introduction

The idea of writing this book came to me after some random observations and reflections. While working in a large academic center, I noticed there needed to be more connection between the artistic-literature-based dimension of medicine, held primarily by ethicists, and the scientific-clinical one, practiced mainly by physicians. After spending quite some time in clinical practice, I had a decent understanding of the scientific foundations of modern medicine. At the same time, I was one of very few among my colleagues who were passionate about the humanistic aspect of medicine or what many scholars refer to as the Art of craftsmanship. After graduating from Internal Medicine training in 2018 and spending a handful of time educating myself about the field of bioethics, I realized that there is so much richness and depth to what we do as a physician beyond the mere fact of understanding physiology and following guidelines. The craft of practicing medicine was not only confined to biomedical healing but also involved learning how to talk to a patient suffering from terminal breast cancer or debilitating arthritis, how to console parents after their son's death in the medical ICU, how to deal with an angry and frustrated family who lost all the trust in the rounding teams yet still hoping that someone could cure their sick patient, and how better to understand the nature of suffering as vulnerable physicians so we can relate more emotionally to our suffering patients.

The clinical practice is, at its core, a relationship between a physician and a patient, a relationship between a sick person who willingly puts this trust in another asking for help. All our clinical duties in daily practice are grounded in this relationship. A good doctor knows and understands her patients. She is the clinician who tries to examine the sickness of her patient in the latter own existence and experience. She is the human who empathizes with her patient and doesn't reduce him to a blocked coronary artery, a COPD case, a type II Carpentier mitral valve pathology, or a pancreatic cancer.

Understanding the break in the existential reality during sickness is pivotal for healing. Healing means restoring a suffering patient's wholeness and re-harmonizing the disturbed relationship between himself and the surrounding world. Unfortunately, one of the unintended consequences of applying modern scientific methodologies in medicine was its fragmentation. As clinicians, we feel and witness this disconnect every day. We refer patients to specialists of different organs to care for the organism. Each patient has a cardiologist, a nephrologist, a pulmonologist, a hematologist, and sometimes many more. And the primary care who is quite often no longer a primary care but rather a gatekeeper and a referring center. A physician who has a twenty-minute follow-up visit to read and understand five complicated specialist notes and two long hospital admissions before trying to update his patient on the latter health progress and clinical care.

I am not trying here to draw a melancholic picture of modern practice, denying the tremendous successes it brought to us. Nor am I suggesting that we should go back to the humoral theory and practice phlebotomy to treat pulmonary edema. Yet I am trying to present some of the benefits and drawbacks of modern practice by taking the reader on a journey of various topics. And although they might seem disconnected at first glance, all the discussed subjects are relevant in one way or another to daily practice.

Part I of this book focuses on the models and challenges of modern practice. It provides a longitudinal examination of the relationship between the physician and the patient. It also discusses the consequences of the commodification of health care, the concept of patient satisfaction, its foundational theories, and some of its main critiques. Moreover, to set the ground for a better understanding of modern practice, the part I sheds light on the weaknesses and strengths of two central models of scientific thinking, namely reductionism and holism. It also examines the roles of rationality and empiricism in clinical practice and why physicians need to understand both schools of thought in what we now refer to as Evidence-Based Medicine. Finally, this part offers a comparative analysis of secular and religious bioethics by emphasizing the role of agapistic ethics in changing the field of medicine from a mere practice to a call and vocation.

On the practical side, when it comes to education and training, it is hard to imagine a medical student, a resident, or even a staff who is not involved in some research activity, whether translational, observational, or a clinical trial. Part II of this book, I focused on some of the practical issues in clinical and non-clinical research. This part examines the principal ethical issues in the research conduct. It illustrates the philosophical and ethical differences between the role of a physician and the role of a scientist, where the two paths intersect and when they do diverge. It also discusses why clinicians should be aware of this intrinsic conflict between the two positions and how it relates to the ethical conduct of research, informed consent, and conflict of interest.

Moreover, in dealing with research, the major advancement in big data and genetic analysis brought us a different kind of challenge and a new sense of unease. Therefore, the second chapter of this part focuses mainly on this issue. It examines the philosophical theories behind human enhancement and new eugenics, discussing the cheers and fears, the hopes and dangers that might confront us soon.

As a clinician interested in advanced heart failure and critical care, it was challenging not to address and put a bright spotlight on these two fields of internal medicine. I dedicated two chapters to discuss some of the major ethical conundrums in cardiovascular medicine. The first explores the various challenges of taking care of patients with advanced heart failure. It offers a narrative analysis of the disease as a chronic sickness affecting many dimensions of patient life and how advanced heart failure can be better viewed through the lens of philosophical anthropology. This chapter also discusses another challenging scenario in cardiovascular care, namely the role of repeat valve replacement surgery in IV drug users who suffers from infective endocarditis, how many surgeries should be offered in case of recidivism, what are some of the theories of addiction, and how can a physician better understand the concept of autonym, futility, and rationing in patients care.

While cardiology continues to be one of the most advanced fields in medicine, these technological and clinical advancements brought us tremendous challenges regarding device management, especially in terminal-ill patients. To address some of those challenges, I dedicated a specific chapter focusing on pacemaker management at the end of life. This chapter analyzes this dilemma from an ontological and phenomenological standpoint and offers a practical guide to approaching patient care in such a clinical scenario. Furthermore, it illustrates the ethical differences between euthanasia, withholding, withdrawing life support, and allowing to die, giving physicians a better understanding of the moral and practical differences between those various notions.

Moreover, concerning end-of-life issues, caring for patients at those critical times requires a unique set of skills and a deep understanding of human suffering. While many of these skills are subconsciously learned in clinical rotations, there needs to be more focus on teaching trainees how to better understand the phenomenon of sickness. There is also little concentration on setting specialized training to teach the non-scientific aspects of patient care, such as human suffering, the role of compassion, the virtues in medicine, and the value of human life. Based on this observation, I thought it would be helpful to dedicate part of this work to examining some of the previously mentioned notions. Therefore, this book's last part addresses many ethical and humanistic notions of end-of-life care. It offers a philosophical analysis of the treatment of CPR, why it differs from other therapies, and how understanding that could help physicians during goals of care conversations. The last part also focuses on the role of compassion and trust in caring for patients. It also deals with some of the challenging scenarios that could face a clinician while addressing end-of-life care, how to respond to unrealistic hopes, and how to better understand the moment from the patient and his lover's side.

In the end, as Professor Edmund Pellegrino said once, "medicine is a moral enterprise." It is an art based on science and a profession guided by ethics. I hope this work will help and slightly enrich the practice of many of my colleagues. I also hope it will build more bridges and start better dialogues between those on the ethics and those on the clinical side. We are all working to serve, and we all need the expertise of each other. Medicine is all about that. It is about that person who could be any of us, needing some of us at one time or another. Medicine is part of who we are, a part of our collective memory, and a wonderful story of how we could relate to each other, love each other, and care about each other at the most difficult times of our lives.

Contents

Part I
Context of Bioethics in Clinical Practice

Chapter 1
Physician—Patient Relationship and the Challenges of Modern Practice

Questions

- What are some of the challenges in delivering holistic care in contemporary patient practice?
- Mention two of the possible conflicts of applying the principle of autonomy in clinical practice?
- Could there be a role for a paternalistic approach to patient's care and what is the relationship between paternalistic approach and the principle of beneficence?

1.1 Introduction

Sometimes, a day at work passes so well, we offer decent help to a suffering person, a colleague brings us a warm cup of coffee, or we receive a beautiful thank you card from a patient's family. And many times, we feel totally burned out and anhedonic, and everything around us in the hospital seems to be going just wrong. Perhaps, that is a common theme in almost all careers; I mean, if you asked around, no one would claim to have awesome workdays all the time. However, when it comes to working in the medical field, things tend to be a little bit more complicated. Practicing medicine, at its core, is quite rewarding. Even when it is not, due to a plentitude of factors, such us, a poor prognosis of a disease, a clash with colleagues, or difficult family dynamics, any health care professional could sense, if he reflected on his work for two minutes, the nobility and sincerity of what he does. Yet despite all this internal goodness of medicine, health care professionals are currently facing many existential career-related crises. We all feel that something went wrong in medicine, but it is tough to precisely pinpoint what it was. Many times, we feel that we are only cogwheels in this giant economic machine called health care industry. This chapter might not address all the causes of this crisis, but it will uncover some of the core values in medicine that I believe got lost in contemporary practice. Medicine, at its core, is a moral enterprise, as Dr. Edmund Pellegrino once said. And perhaps what

© The Author(s), under exclusive license to Springer Nature Switzerland AG 2023
S. Toro, *Introduction to Clinical Ethics: Perspectives from a Physician Bioethicist*,
https://doi.org/10.1007/978-3-031-30804-8_1

is really challenging physicians these days is primarily an ethical demand. From a Shaman to a provider. From a father of a sick man to a customer service doctor aiming to satisfy a patient, here a physician is writing and trying to shed a beam of light on a few meaningful virtues in medical practice, hoping that those virtues might offer some guidance and hope for the coming days ahead.

1.2   Physicians, Patients and Illnesses

1.2.1   From Shaman to Provider, a Historical Review

Since the beginning of time, man has suffered from various maladies. In an attempt to face suffering, he questioned, reflected, practiced, and learned. A mutual relationship between a man in need and another who can offer help was born.[1] This relation, which is called today the Physician—Patient relationship, was and still is the central pillar of medical morality.[2] Throughout history, this relationship was primarily influenced by two dominant forces, the degree of medical knowledge and the socio-economic scene of the particular era in which the relationship was found.[3] For instance, in pre-literature and among primitive people, doctors were "Shamans" or "Witch Doctors." These healers practiced medicine mainly through rituals and magic. Diseases were thought to be caused by supernatural powers of demons, and Shamans were exclusively experienced in dealing with these supernatural forces.[4] In ancient Mesopotamia, we find a similar relation between the Gods and disease; in Hammurabi's Code, for example, a link is disclosed between "Sins" and "Illness".[5] Interestingly though, during the same era, and away from metaphysical healing practices, we also start noticing the budding of the culture of craftsmen. The "asu", for example, was not a priest but rather more of a primitive herbalist, who offered help to the sick by physical practice rather than the common spiritual ways.[5]

Moving chronologically, the Greco-Roman era witnessed two significant signs of change in medical culture: (1) First, medicine became more empirical, and craftsmen focused more on explaining diseases outside the category of mythology and religion. For instance, they turned into philosophical rationalization, trying to find common ground to understand the nature of man.[6] (2) Second, new concepts started to shape and influence the physician–patient relationship. In particular, the Hippocratic Oath which established a code of conduct for physicians and a statement of rights for patients.[3] Also, there was an overall change in the attitude of society toward the sick. This change was revealed mainly in the rise of the concept of philanthropy and the Stoical idea of the universal kindship of humankind.[7]

Later on, Christianity came to the world as a religion of healing. It made a revolutionary change to how society looked to the sick and poor.[8] As a practical dogma, Christianity emphasized "Agape," the unconditioned, unbounded love, that resembled the sacrificial love of God to humankind.[9] From the early Christian centuries toward renaissance, the reader can notice again that the practice of medicine involved two types of physicians. On one side, monks practiced their physician sort of role as an extension to the clerical duties. On the other side, secular craftsmen practiced medicine as a career to make a living.[10] It is also worth mentioning a common theme regarding the general attitude towards physicians in all those eras that we reviewed so far. From early history, physicians had to struggle to gain a patient's trust—the image of physicians contrasted between an "ideal" and "incompetent" was quite common. Quotes like "physician, heal yourself "are often found in medieval and renaissance literature.[11] Such an attitude is not surprising, as most of the treatments were nonspecific, primitive, neither effective and nor based on any robust scientific framework.[12]

After the eighteenth-century significant advancements were achieved in the field of physiology, microbiology, and surgery.[13] A better understanding of the pathology of various diseases enabled physicians to deliver more targeted and effective care. Hospitals were built, and physicians gained profound experience from practicing and getting exposed to a variety of patients.[14] During these times, the paternalistic model of relationship between a patient and a doctor was the dominant model, and it continued to prevail till mid twentieth century. After the 1950s, significant social and economic changes started to influence the conduct of medical practice. Power and authority shifted gradually from physicians to patients with more emphasis on patients' autonomy. Doctors were no longer the parents in the traditional paternalistic model. The principle of beneficence lost its place as the central pillar in the physician–patient relationship.[15] These changes were driven mainly by the culture democratization and the rise patient's rights movements.[16] Added to that, the easy accessibility to medical knowledge, which decreased the knowledge gap between physicians and patients and enabled the latter to question doctors' judgments.[17]

Economic factors also played a significant role in the scene, especially in the United States. The increasing cost of health care and the deterioration of economic growth, especially after the 1970s, escalated the tension between payers and providers. Businesses started to self-insure their employees rather than paying insurance companies to decrease the cost. Large health maintenance organizations grew, expanded, and swallowed up small private practice offices.[18] As a result, medical practice became a commodity and required necessary help from experts in the field of financial management. All these changes interfered directly with the privacy of the patient-physician relationship, which became at the end a relationship between a consumer who is shopping for health care and a provider who is employed to provide it under the organization's conditions.[13]

1.2.2 Disease and Illness

Many theories have been suggested by scholars on how to define a disease. Each of these theories partially succeeds in explaining the concept yet fails to provide a solid theoretical background to generalize the definition. The first of these theories is the Ontological theory, described by Virchow in 1895.[19]According to this theory, diseases are separate entities that affect or invade the human body causing sickness. The germ theory provides a good example of this framework. Diseases are caused by bacteria, viruses or parasites that invade the host causing a variety of pathologic manifestations. While the germ theory explains many different infectious pathologies, it fails to provide an explanation for many other known diseases, such as coronary artery disease, alcoholic pancreatitis, and Rheumatoid arthritis.[20]

The second theory is the physiological theory. Chronologically it precedes the Ontological theory of disease. In fact, the same Virchow, initially in 1847, refused to consider diseases to be separate existing entities. The Humoral theory of disease, which dominated the medical world till the late nineteenth century, provides a good example of the antiquity of the physiological theory. Health was defined by the balance between four body fluids (phlegm, blood, black bile, and yellow bile) and the surrounding environment. Blood was hot and wet, resembling air, on the other side; phlegm like water was wet and cold, yellow bile corresponds to fire; being hot and dry, black bile was linked to earth by being cold and dry. Diseases resulted from the disturbance in the equilibrium of these four humors. For example, in the winter, bronchitis was thought to be the result of the excessive amount of phlegm relative to other humors. The job of the physician was to try to reverse the disease state by restoring the balance between different fluids.[21] However, the physiological theory of disease is not confined only to the ancient humoral theory. With the significant progress in physics and chemistry, physicians started to understand further how the human body functions. The science of physiology emerged to explain the mechanisms of various organs function. Normality was defined based on a statistical measurement of Gaussian distributions. For example, normal systolic blood pressure ranges from 120 to 140 mm/Hg. Readings above or below that range might indicate a disease state.[22] Broose added a functional component to it, stating that diseases should also interrupt or interfere with different human functions. He named his theory the Biostatic theory.[23] According to him, the teleological aspect represents the ability of a natural class to survive and reproduce, a concept, as we will discuss later, was not enough according to Sulmasy to define diseases from a teleological standpoint.[24]

The next theory is the evolutionary theory. Based on this particular approach, diseases result from a maladaptive evolution of a given species. For example, tuberculosis results from an immune system' failure to fight mycobacteria. Morning sickness of pregnancy is an adaptive physiological mechanism to prevent the mother from ingesting toxins during the first trimester when most of the fetus organs are formed.[25] Fever is also an evolutionary mechanism developed by the body to fight various infections. However, critics of the evolutionary theory point out that while evolutionary biologists are mainly concerned with reproductive fitness, doctors care about the overall well-being of a patient. Coughing reflex as an adaptive mechanism might help a patient to expel pathogens from the chest, but coughing is also unpleasant, irritating and sometimes painful. Many patients visit the primary care clinic to get a prescription for cough suppressants; explaining the symptom to them from an evolutionary standpoint might be interesting to hear but not practical from a medical standpoint.[22] And examples of similar cases are too plenty to be mentioned.

The last theory in understanding diseases is the genetic theory. According to this theory, different illnesses are caused by various genetic changes and mutations. They can follow either the classic Mendelian inheritance like Sickle cell disease, the polygenetic inherence like lung cancer or they can be a manifestation of the mitochondrial genetic disorders like MELAS and MERRF.[26]

Before moving in the discussion toward explaining the concept of illness, another approach in disease definition is worth mentioning. In his paper *Disease and Natural Kinds*, Sulmasy proposed the following points: (1) a disease is a set of circumstances that affect a natural kind. (2) The disease has to affect at least one natural kind. (3) Some diseases might affect many natural kinds. For example, myasthenia gravis affects humans and dogs. (4) Natural kinds have dispositional properties that determine their teleology. (5) A disease is a disturbance in these properties (law-like principle) that interfere with the natural kind ability to flourish. Reproductive function is not enough for determining the human natural kind flourishing. Preserving a reproductive and survival function—from a teleological standpoint- might be enough for a simple organism like amoeba but not for a human being. A disease might not manifest as a pattern of disturbance in some individuals of a natural kind. However, for a disturbance to be considered a disease it should have been at least manifested before in some other individuals. For example, some people die with silent localized prostate cancer that did not interfere with their flourishing, however, prostate cancer had affected a multitude of other patients and interfered with their life flourishing process. (24)

Illness, on the other side, is a broader concept. It does not focus only on providing a theoretical framework on the nature of sickness and the origin of a disease. On the contrary, the concept of Illness carries a social, emotional, and cultural dimensions.[27] It describes the unique experience of a particular patient with his or her disease. When doctors, for example, explain the pathogenicity of Hepatitis B for a patient, they express their general knowledge about the virus, the incubation period, the symptoms, and the possible risk of having hepatocellular carcinoma. They disclose how Hepatitis B virus behaves in general and what might patients experience based on surveys and statistical reviews. To what extent and how a particular patient will encounter the disease is something hard to predict. The social and emotional dimensions of such experience are much more complicated.[28] Take another example; Pneumococcus pneumonia is one of many bacteria responsible for causing lobar pneumonia. Restricting the pathogenicity of pneumococcus pneumonia only to the bacterium might sometimes be too simplistic. Lack of heating, poverty, malnutrition, life stressors, might all be precipitating factors of the sickness.[29] Similarly, in depression, a disturbance occurs in the concentration of some neurotransmitters in the brain, but no one believes that depression is solely and simply caused by such disturbance.[30]

Moreover, the meaning of suffering comes from two main aspects; the importance and the significance of a disease, and the values that a particular patient holds. Diseases are perceived at multidimensional levels. For example, for patient X, chemotherapy in the scientific sense is a medication that suppresses cancer by a specific mechanism. Emotionally in the consciousness of the patient, the treatment means the weeks of nausea, fatigue, fevers, and mouth sores. Chemotherapy might also represent a spiritual reflection, God's help, and sincere hope for recovery and wellbeing.[31] Reducing the definition of heath only to the absence of a disease state, and the concept of disease strictly to the biomedical sense had negatively affected patients care and significantly contributed to the contemporary crisis in health care. (27)

1.3 Models of Relationship Between Physicians and Patients

1.3.1 Beneficence Based Model

In this model, the concept of patient's benefit becomes the cornerstone of the relationship between the physician and the patient. Before proceeding in the discussion of this model, it is essential to make a significant differentiation. Beneficence based model is not the same as the paternalistic model. In the latter, the physician assumes the patient's values and act like a father or priest in his or her benefit.[32] In extreme forms of paternalism, patients are expected to comply and obey- without even questioning—the physician's recommendations.[33] Strong paternalism violates patients' autonomy and hurts the patients.[34] On the contrary, in the beneficence based model the physician does not impose his opinion; instead, he helps a vulnerable patient to reach a consensus of values through mutual dialogue.

There are three components of in the relationship according to the beneficence-based model: (1) the fact of the patient illness, (2) the act of the profession, (3) and the act of medicine.[35] Regarding the first component: an illness affects daily life, whether it is a simple Flu or a broken leg.[36] To the patient, the experience of the disease is not merely the manifestations of its symptoms, but rather the meaning of the perceived malady on social, psychological, emotional and spiritual levels.[37] A new cancer diagnosis in a young father of three kids will carry a burden not only on himself but also his family and relatives. It will affect his personhood and individuation.[38] The healthy body, which was not recognized before the occurrence of the disease, becomes the central concern, as the human existence is threatened.[39]

The second element in the relationship is the fact of the profession. To *profiteri* is to declare aloud. It is a promise given by the physician to a person seeking help. This promise requires two conditions, first that the physician is competent and possesses the required knowledge. Second, that he will use this knowledge for the patient's good; that he will value what is important to the patient, and what defines a good life for him or her. The last element is the act of medicine, the act of healing the individual as a whole, helping him or her to restore life harmony. Not only focusing on specific organ dysfunction but instead taking a holistic approach in delivering care.[26]

Mallia suggested a phenomenological approach in examining the relationship between patients and physicians. He referred to Heidegger description of the existential fact of "concern."[40] The physician enters the relationship to offer help and care. The patient trusts the physician to act in his or her best interest. One enters to deliver care and the other to receive it. The concept of mutual beneficence between a physician seeking an internal good, and a patient receiving care is the central theme of the relationship.[41] Based on this approach, Non-maleficence is a consequence of beneficence grounded on trust.

Despite medicine becoming more scientific and more evidence-based, physicians continued to struggle in decision-making. According to Mallia, non-maleficence is rooted in the trust of the physician to act virtuously in the light of medical uncertainties.[42] Autonomy, on the other hand, was viewed a phenomenon of libertarian philosophies. It was a result of enlightenment ideas, Kantian ethics, and later, the civil rights movement in the 1960s.[43] Per Mallia, autonomy has no roots in medicine, and what is now discussed as autonomy has been viewed before under the concept of beneficence.[44]

1.3.2 Autonomy Based Model

Now, let us move to discuss two related Autonomy based models of physician-patient relationship: the business-based model and contract-based model. The business model, according to Pellegrino, views healthcare as a commodity. Patients shop around in the marketplace of healthcare as consumers. Doctors offer their services to the patients as providers. The ethics of the relationship becomes mostly a form of good business. Physicians are expected to offer a good product, and patients have to pay the bill to get the service.[45] Some scholars suggested that consumerism in medicine provided a chance for patients to challenge doctors' authority, as they can negotiate and have a voice in the service provided.[46] However, the roots of such a change in the relationship does not extend solely to the movements against paternalism. It is part of the overall consumerism movement that started in the 1960s, empowered by the legislation of consumer rights. It is also a result of the gradual shift in practice from curative to preventive medicine, wherein the latter the patient has to be persuaded and encouraged to get regular checkups and specific screening tests.[47] In the business-based model, all parties come to a free market as equals to bargain. The central value is economical. The patient tries to get the service at the best price, and the physician searches for the highest possible income. One example is the shortage of primary care physicians which is mostly based on economic reasons of lesser income compared to specialists.

Many scholars have criticized the business model for different reasons. First, it is not accurate to presume that the relationship between physicians and patients is egalitarian. On the contrary, patients seek health care burdened by sickness and vulnerability, which places them in a weaker position. Additionally, despite the overall improvement in medical knowledge in society, the knowledge gap is still present, and it is in favor of the more knowledgeable, namely the physician. Second, the competitive nature of the marketplace brings out the worst of human behaviors.[48] Third, the very fact that what is being purchased is not an external product but rather the patient's health. In other words, the patient in this model is precisely the market commodity.

The contract model is an extension of the business model in a formalized framework.[32] Like the business model, it was born as a response to hard paternalism and in favor of the patient's autonomy.[49] Different forms of contractual models have been suggested. Some scholars argued in favor of a legal-commercial contract. Others preferred a form of a social one, similar to traditional religious or marriage contract.[50] According to William May, the contractual model is very appealing for many reasons. First, it limits the physician's authoritarian power and consequently the tendency to abuse it. Second, it concentrates more on informed consent rather than trust. Third, it specifies the duties, conditions, rights of both the patient and the physician. Lastly, it provides legal protection for both parties.[51] Brody elaborated on other pivotal features of the commercial contractual model that neither the physician nor the patient is obliged to enter into the contract.[52] Treating a sick patient is not mandatory on the physician. Seeking health care and treatment are not obligatory on the patient. Later on, Brody shifted his position and argued in favor of another model, the stationary model. In this model of relationship both the physician and the patient have obligations. The former treats the patient well for a reasonable fee, and the latter seeks help to maintain the highest possible own health. Many critics raised concerns against this model of relationship. Some of those concerns included the tendency of this model to encourage the practice of defensive medicine and to promote mistrust. Additionally, this model of relationship reduces the professional obligations to the minimal limits, based only on what is written in the contract, which is indeed unpractical in the highly complex clinical practice.

1.4 Patient Satisfaction

1.4.1 The Rise of the Concept

Since the late 1970s, a lot of effort has been put into studying the concept of patient satisfaction. Many scholars viewed the rise of this phenomenon as a part of overall movements toward consumerism in medicine.[53] The democratization of health care, the tendency to challenge physicians' authority, and the increasing focus of re-defining health care delivery in the light of consumer rights, were all contributing factors to this change.[54] This consumerism movement led to a gradual substitution of terminologies in clinical practice, where physicians became providers, and patients became consumers.[55] The aim of this fundamental change in the relationship was to provide patients with more dignity and rights to stand against classical paternalism.

Consequently, malpractice, lawsuits, noncompliance with physicians' recommendations all increased drastically.[56] Overall, there was a tendency to empower patients and involve them more in the health care delivery. Another significant factor in the emergence of patient satisfaction was the increasing cost of health care. Caused by the gradual shift from acute to chronic pathologies along with the significant progress and advancements in medical technology. Consequently, consumers began to criticize the high costs of health care delivery, claiming rights to be involved in evaluating it.[57] As a result, the effectiveness of medical care started to be measured not only based on clinical criteria but also on economic factors. The philosophy of "quality" emerged and was encouraged by health care organizations. Governments in the west started to promote a competitive health care market policy, and hospitals had to ensure achieving high patient satisfaction levels. For example, Medicare hospital value-based program (VBP) paid hospitals 1.4 billion dollars in 2015, 30% of the VBP incentives were based on HCAPS.[58] Despite the significant contemporary focus on the concept of patient satisfaction, there are still vast disagreements about its definition, reliability, and validity in health care delivery.[59]

1.4.2 The Significance of Patient Satisfaction

Someone can view the significance of patient satisfaction measurements under two main categories: (1) through its direct effect on patients and physicians, and (2) by its role in the general evaluation of health care delivery. Higher patient satisfaction scores are linked with increased compliance with medications and treatment plans.[60] Increase compliance leads to better outcomes and quality care.[61] Satisfied patients are also less likely to complain against health care organizations and hospitals.[62] In contrast, lower patient satisfaction scores are associated with increased risk of malpractice lawsuits. In a study of 353 physicians in The Brigham and Women Hospital in Boston, the rate of malpractice lawsuits was 110% higher against physicians with bottom tertile compared to those in the top satisfaction survey ratings.[63] Patients are far less likely to file complaints against caring and friendly physicians, and it is well known that good communication and personal caring are essential determinants of patient satisfaction.[64] On an organizational level, high patient satisfaction leads to greater patient loyalty, since satisfied patients are more likely to continue to receive their care in a given organization or hospital.[65] Reichheld well demonstrated the economic effect of such behaviors. High patient loyalty leads to increase profitability. The rise of patient retention by 5% increases the value of the customer by 25–100%.[66] Additionally, satisfied patients are more likely to refer their relatives, friends, and the people they know to get treatment at the facilities where they experienced good care. On the contrary, patients who had bad experiences with an organization will also communicate their dissatisfaction and frustration.[67] This chain of communication has an immense effect on the performance and profits of institutions. Lastly, patient satisfaction research also helped in evaluating health care delivery from patients' perspectives.[68] Third-party payers and governments were interested in investigating how patients perceive health care delivery. Satisfaction research identified many defects in the health care system and led to the implementation of many amendments in its conduct. For example, in 1978, the Royal Commission in Britain conducted a large survey to identify problems related to health care delivery in complex organizations. The results showed that two-thirds of patients' dissatisfaction in the hospital could have been eliminated by waking up patients in a more reasonable time and updating them about the progress of their hospitalization course.[60]

1.5   Definition and Critique

1.5.1   Theories of Patient Satisfaction

The expectancy-value theory of Linder-Pelz defines patient satisfaction as "a positive evaluation of distinct dimensions of health care." Satisfaction is an attitude toward received care. Attitude is not only a perception of what has been received but also an evaluative process, an assessment of the multiple dimensions of the object. Patients have a specific set of expectations toward what they should get when they seek medical care. This collection of previously set expectations significantly influences whether patients will be satisfied with the delivered care or not. If patient desires and expectations are not fulfilled during the clinical encounter, dissatisfaction with care will follow.[69] Studies have shown that patients with lower expectations tend to be more satisfied. Interestingly, increasing quality of care raises patients' expectations and might result in an overall decrease in patient satisfaction.[70] Stimson and Webb suggested three components of expectations: background, interaction, and action. Background expectations are a set of beliefs toward health care delivery.[71] Patients derive them partially from intrapersonal comparison with their previous experiences and partially from the interpersonal comparison with other individuals. These expectations can differ between societies, ethnic groups, various countries, and they significantly influence patients' evaluation of the received care. The interactive component of expectations deals with what patients expect during the clinical encounter: the bedside manners of the physician, the communication skills, and the technical and non-technical aspects of the clinical visit. The last component deals with the provided actions or the treatment plans. With the current wide availability of medical information, patients usually research their illnesses before visiting the physician. Sometimes they expect a specific treatment or procedure. If the doctor did not provide the anticipated action, dissatisfaction with the service would result.

Fitzpatrick argued that satisfaction should be viewed in three different models.[72] The first one, "the need for the familiar," which converges with the previously discussed theory as it considers satisfaction an attitude influenced by common social determinants.

The second model, "The goals of help-seeking" suggests that patients' satisfaction is largely dependent on the outcome of the visit. Fitzpatrick argues that patients' expectations are tentative, provisional, and not fixed. What matters for them at the end is whether their illness is cured, improved or not. In his qualitative study on Neurological consultation in eleven NHS hospitals in Britain, most patients had a low set of expectations from the consultant. Patients expressed these expectations as "possibilities" of what might happen. Fitzpatrick pointed out that the different experiences of the illness (which in this study was s headache) influenced what patients expected from the doctor. Patients who were concerned about malignancy needed reassurance. Those who failed multiple treatments before expected advanced therapies. Patients who had recurrent spells of disabling headaches were mostly hoping for prevention. When consultants did not capture which dimension of headaches concerned different patients, disappointment was noticed in the post-visit evaluation.[73] The last model suggested by Fitzpatrick "the affective behaviour" is largely based on the research of Ben-Sira on primary care visits. This model proposes that patients' satisfaction is mainly dependent on the mode of physician action rather than the content. Ben-Sira argued that patients are usually burdened by significant anxiety when seeking medical care. The degree of anxiety and uncertainty increases with the severity and seriousness of the illness. Patient satisfaction will be largely influenced by the ability of the physician to reassure the patient and relief his/her anxiety rather than the content and outcome of the visit. In his study on 1892 patients in Israel, Ben-Sira found a strong correlation coefficient (0.68–0.86) between satisfaction and physician's affective behavior (devotion, time, and interest). He argued that since patients are incapable of judging the instrumental aspect of the delivered care, their satisfaction will be mostly dependent on the affective and emotional behavior of the clinician.[74]

1.5.2 Critique to the Concept

Numerous studies have been published since the 1970s on patients' satisfaction. Literature review reveals that despite the abundance of published materials, many scholars continue to question the validity of the concept. Generally, the criticism against patient satisfaction can be viewed in three different areas. The first deals with conceptualization, the second with the methodologies, and the third with the outcomes.

Until now, no unified consensus on the theoretical framework of patient satisfaction has been reached. Many theories were proposed, each of them has its own defects.[75] In regards to the methodological concerns, Locker and Dunt showed that patients tend to be less critical when they are asked open- ended questions.[76] Also, patient's dependence and need for future medical services might have a significant influence of their evaluation of the clinical encounter.[77] Furthermore, patients tend to be less critical of the technical aspect of health care as many believe they lack the competence to evaluate that esoteric aspect.[78] Another important point is that patients' evaluations tend to be unstable, unfixed and greatly influenced by their emotions. Sometimes this "halo effect," a single striking reason or impression, can significantly shape the overall final evaluation.[79] In addition to the previously mentioned methodological problems, scoring system remains a key problem in satisfaction research. Multiple scales have been suggested without a clear consensus on their validity and reliability. The lack of a standardized instrument for measurement makes comparing these scales quite challenging. In a review of 195 articles published in 1994, Sitzia showed that only 46% of them reported some validity and reliability. Also, 80% of the studies reviewed in his work used new instruments for measuring patient satisfaction, 60% of those did not discuss the validity and reliability of the new measurement scales.[80] Another problem with satisfaction research is the individual response rate. In another review, Sitzia showed that only 48% of the studies report the response rate, and the response rate of (face to face) recruitment was higher compared to mail surveys and data collection (76.7% vs. 67%). Sitzia pointed out that many researchers in the patient satisfaction field do not take this response bias quite seriously.[81]

Finally, the last critique of the concept of satisfaction concerns the outcomes. Advocates for patient satisfaction assert that increasing patients' satisfaction leads to increased patients' loyalty and increased compliance with treatments, which will eventually lead to better financial and clinical outcomes. However, the evidence is not quite clear regarding this assumption. For example, in a large retrospective cohort study of 51 thousand patients, respondents with the highest satisfaction quartile compared to those with the lowest one had increased odds for inpatient admissions (OR 1.12; 95% CI, 1.02–1.23); they also had higher total expenditure and higher mortality (adjusted hazard ratio, 1.26; 95% CI, 1.05–1.53).[82] In many hospitals and health care organizations, physicians' compensation is strongly determined by patients' satisfaction scores. Physicians working in such institutions tend to order discretionary and sometimes unnecessary tests to meet their patients' expectations. High expenditure is associated with ordering more investigations, specialists' referrals and it was shown to be linked to higher mortality rates.[83]

1.6 Achieving the Goals of Medicine

The last section of this chapter will try to shed the light on the importance of reaching a consensus in defining the main goal of medicine. The goal presented in the healing relationship between a patient in need and a physician who promise to help. The first passage will discuss the essentiality of the concept of virtue in clinical practice. The second passage will present some of the contemporary changes and challenges in clinical practice and it will mainly focus on the risks associated with commodification of health care.

1.6.1   Virtues in Medicine

Aristotle defines virtue as a "state of character "that makes a human a good person and allows him to do his work well.[84] For an act to be virtous, it has to come from a virtues person. It is not enough for a physician, for example, to perform a technical procedure successfully, if his intentions were not aimed primarily for the good of his patient.[85] A physician who does not have a vitreous character cannot always be trusted.

Throughout history, the concept of virtues was a topic of extensive debate. It witnessed significant changes since the early arguments of Plato, Socrates, and Aristotle. Providing a detailed discussion of the major philosophical theories of virtues is out of the scope of this chapter. However, it is worth briefly summarizing the main changes that occurred in the general understanding of the concept.

For Aristotle, virtues were of two main kinds; the intellectual virtues of Sophia, science, reasoning, phronesis, and the moral virtues of fortitude, justice, temperance, and wisdom.[86] Humans gain moral virtues through practice. The etymology of the world Ethika is close to the world Ethos (habit). Good character and good works will enable a man to achieve Eudaimonia, the state of human flourishing in life. Stoic philosophers also believed in the law of human nature. A man should act in accordance with God, being benevolent and free of disturbing passions. In the medieval period, Thomas Aquinas added the Christian virtues of Faith, Hope, and Charity to the classical cardinal virtues.[87] However, the teleology of virtues man for Aquinas was not the human flourishing but rather a spiritual destination aimed at the union with God.[88] Following the medieval period, significant changes occurred in the world view of virtues through the emergence of the Kantian ethics of duties and consequentialist ethics in the Utilitarian theories. Those theories shaped and significantly influenced the contemporary focus on principle-based rather than virtue-based ethics.[89]

However, the question that comes into our minds is, why do we need virtues? The answer to this question is not a straightforward one. As much as it is challenging to find a common consensus in defining virtues, it is also equally challenging to point out exactly how they help humans practically. First, let us start by discussing the concept of patient's good, the importance of health as a value in human life, and what could be the main goal of medicine?

Heath is a valuable state in humans' life. On a societal level, it is one of the major concerns for any community. Access, availability, cost, and many other areas of concern regarding healthcare are routinely discussed on TV channels, newspapers, and journals.[90] Health is also a valuable state on the individual level. Ontologically, the body prefers health; it tries to repair itself whenever an insult occurs. Health is also a value when a patient subjectively feels that something is going wrong and seek medical help. It is also a value in the presence or absence of objective signs that can be pointed out to a patient when others perceive his sickness. When the patient becomes sick, he enters a relationship with a physician. Both of the parties in this relationship value health. One enters seeking restoration of a healthy state, and the other promising to offer it.[91] The goal of medicine is presented in this mutual valuation of health and through the relationship between a physician and a patient. However, the questions to be asked here: is health always an absolute value? What about a Jehovah's witness who commits to refuse blood transfusion knowing that he might die? To further clarify, we have to discuss the concept of patient's good more thoroughly.

Pellegrino demonstrated the good of the patient in four main levels. The first is a biomedical good; the end that is achieved through the scientific knowledge and the technical abilities of physicians—for example, reducing pain, opening an occluded coronary artery or removing a ureteral stone.[92] The second level is the patient's judgment of what is good for him or her. It is a value judgment of whether prolonging life or achieving a particular medical goal takes priority. It is the balancing process between what can be achieved and what might be lost. For example, proceeding with chemotherapy or taking the risk of a major surgery. The third level of patient good focuses on the patient's free will and his or her capabilities of acting as a free agent; being mentally and physically able to make own decisions.[93] Finally, the last level deals with the ultimate value of a person in his or her life. The *Summum Bonum*, whether it is metaphysically or physically determined. For Pellegrino, the goodness of free will and ultimate value takes priority over patient's value judgment and biomedical good. Refusing to respect a JW decision to refuse transfusion is a violation of his or her *Summum Bonum*.[94]

Nevertheless, the question of why we need virtues in our life is a relevant question. As C. S. Lewis brilliantly expressed it, laws are not enough to make good men. Every time people come up with what seems to be a resilient system of principles; someone, somehow finds a way to corrupt and take advantage of it.[95] Following the same conclusion, virtues are essential to any physician's life. By them, she keeps the moral values of the career and protects herself from exploiting the advantages that it brings. To illustrate this concept, some examples of essential virtues of a good physician will be given.

Fidelity in Trust is one of the most essential virtues in medicine. It is also equally essential for any society to function. Trust involves contingency and reduces complexity in the community. Without Trust, we can do no action. For a person to function, he has to trust the competence and the intentions of the people around him. For example, when we go on a flight, we don't question the competence of the pilot or double-check if he knows how to fly a plane. We simply trust the system, airline accountability, and the airport accumulated experience. However, when it comes to the clinical encounter, we rarely, if ever, rely on the system blindly. In medical practice, when a patient enters the relationship with a physician, he trusts that the latter will not exploit his vulnerability. He trusts the doctor to try to restore his health, reduce his suffering and respect his values. Also, he wants to be sure that his physician is honest about what he can and cannot offer in his care. Interestingly Trust, in this context intersects with honesty and humility, another two essential virtues in a good physician.[96]

The second essential virtue to discuss here is compassion. The literal meaning of the word, to *compati*, to suffer with. Compassion is not only a sentiment but a virtue with a moral dimension. The physician cannot understand the illness of a patient if she did not share her sufferings. Compassion is different from pity and mercy, where superiority and condescendence are practiced over another fellow human. It is not empathy in its imaginative form, as commonly said "putting oneself in another's shoes." It is also not sympathy, where a general sharing of feelings might occur rather than specifically suffering. Compassion as virtue carries an intellectual element. It sometimes necessitates *epoche*, when the physician has to slightly withdrawal from co-suffering with a patient so she does not lose her clinical judgment.[97] Here as we will discuss later, prudence comes as an essential virtue to guide the application of compassion.

Fortitude is another necessary but rarely discussed virtue at the present time. As said by C. S. Lewis "Courage is not simply one of the virtues but the form of every virtue at the testing point."[98] It is the moral fortitude to act courageously in the face of adversity and uncertainty. It is the virtue that a physician holds on when he risks his life in taking care of people with contagious diseases. It is the courage to operate on someone who has HIV or to take care of wounded people in a field hospital at war zones. For any physician to function with the ethical requirements of the career, a certain degree of fortitude is required. Fortitude is essential for the sustainability of the moral acts of a free agent. Unfortunately, the depersonalization of our society and the general requirement for a physician to function in an environment of HMOs, have significantly limited the capacity of fortitude to flourish. The role of the physician is now strictly confined within a system of practice, in which an incongruent act might result in isolation, unacceptance, and in worst cases, losing the career.[99]

Lastly, and before moving into the final section of this paper, it is essential to comment on Prudence, a cardinal virtue in the life of any human. Prudence is the practical wisdom that guides the person to act virtuously in the light of uncertainty. For Aristotle, Prudence, or what was called before Phronesis, is the virtue that links and orchestrate other intellectual and moral virtues. It is what guides a person to find the mean between recklessness and cowardice and between secrecy and loquacity. It is the wisdom that the physician determines through the degree of compassion needed to help a patient without blurring out the medical judgment. It is also the wisdom that balances the physician's loyalty and commitment to his career and his duties toward his friends, family, and beloved ones.[100]

1.6.2   The Risks of Commodification of Health Care

In the last few decades, a constellation of social, political, and economic events led to fundamental changes in the general understanding of health care. The wide emergence of managed health care has promoted a different ethical vision to medical practice. New terminologies and concepts were added from the business world to the traditional medical literature. In particular, the effects of these changes were manifested on three different levels; the doctor who has become a provider, the patient who has become a consumer, and the main telos of health care which was changed into a commodity.

To answer the question of whether health care should be viewed as a commodity, a discussion and illustration of the consequences of such a vision will be provided. To start, let us first examine the changes that occurred to the physician and how these changes affected the traditional understanding of the physician–patient relationship. In a world of managed care, physicians are marketed as providers, available for patients to purchase in a free shop. Physicians become purveyors and traders for their employers; primary care doctors become resources to feed the system, and Specialists become a mean for high wedge procedures and interventions. As a result, the loyalty of the physician is now fragmented between patients, insurance companies, and employers.[101] Physicians are encouraged to be more efficient, to see more patients, and order less expensive tests. Patients' needs are no longer placed first in this hierarchy of goals. Additionally, the traditional philosophy of individual-focused care is replaced by guidelines based and quality outcomes practice that treats all patients as equal consumers.[102]

On the other side, patients become consumers of health care; insured lives to be traded between organizations. The intimate relationship with the physicians is destroyed by losing continuity. Once the encounter is done, everything is over; the doctor, in this case, resembled a cashier in a grocery store. The provider-consumer interaction doesn't necessitate a relation outside the service provided. In a free market, a patient who cannot pay is considered a bad customer in the system. Chronically ill, debilitated patients are also not preferred consumers, as they have high demands, need more time, and are more costly to the system. In a world of health care managed through business ethics, there is no guarantee for any form of altruism. If a patient was unfortunate and made the wrong choice, he will be left alone to suffer the consequences of his choice.[103]

Lastly, if health care is a product and a commodity, then the question that comes to our minds is who owns it? Is health care for the providing institution a bar of chocolate, or a new car, that a patient buys and enjoys with satisfaction? In the literal sense, a notion of fungibility is present in defining commodity. However, given the fact of illnesses with all the nuance they entail, and the way patients respond differently to treatment options, it is undoubtedly challenging to compare health with chocolate bars and various cars. Neither physicians nor health care organizations own the medical knowledge. It is a part of our shared human heritage. It is the fund of facts of accumulated experiences that humans have stewardship at. By becoming a physician, the person declares aloud his commitment to the wellbeing of humans and to help those who are suffering. Physicians are responsible in front of the human community to use the medical knowledge to heal and to restore the flourishing state of human beings.[104]

1.7 Summary

In summary, health care should not be viewed as a commodity. Throughout history, the ethical heritage of medical practice was kept and preserved by vitreous physicians who committed their lives for the service of the patients and the community. The healing relationship has always been the core of medical practice. It is the relationship that aims for the patient's good, to heal and to reduce suffering. The recent changes in the way of practicing medicine are alarming. It had negatively affected both physicians and patients in a variety of aspects. Holding onto the traditional ethics and virtues of medicine is essential to preserve the goal of the career in the light of contemporary challenges.

Thought Box (I)
In 2013, in Aleppo-Syria, I was doing part of my rotations in one of the largest cardiovascular centers in the city. Those days were indeed not the most pleasant and easy days, neither from the social and basic aspects of daily life nor concerning its psychological and safety elements. I still hear the sound of the electric generator outside the hospital and the random snipers firing in the dark streets. I do remember the faces of terrified soldiers and the screaming in the emergency rooms. And I still do remember some of the interactions with my patients. At that time, when asked, I used to do some house visits to my neighborhood patients. I had a small bag containing a blood pressure cuff, a stethoscope, some intramuscular injections of anti-nausea, antipyretic medications, IV fluids, and other basic medical equipment. It is pretty challenging to compare the relationship between doctors and patients on that side of the world with the United States. Yet I can still feel the difference when I write my long H&Ps in EMRs and after I take care of an international patient every now and then here in Houston.

When it comes to the physician–patient relationship, the paternalistic model is still the predominant model in many non-western countries. For example, most of the patients in Syria don't argue with their physicians about the treatment plan. Shared decision-making is rarely if every part of the clinical encounter. The doctor's word is extremely trusted at most times, and there is almost invisible religious authority guiding the relationship between a doctor and the patient. Patient satisfaction has a different manifestation. It is reflected socially and economically, as good doctors will get more referrals and more patients. When patients are happy with the care they receive, they tend to refer their friends and family members to a specific doctor, which means more business and fame to the latter. Although a financial component is present in the relationship, patients don't like financially driven physicians. There is a bad connotation when money and business get attached to patient care.

Compared to the practice here in the United States, the decision-making process tends to be more egalitarian. Patients have a sense of autonomy over their bodies and want to say a word in their care. A Doctor rarely has the intrinsic guaranteed trust just because he or she is a doctor. The clinician is often challenged with articles and information and convincing a patient to take a medication is not always quite easy.

From a patient satisfaction standpoint, a broad spectrum of factors plays a role. For example, the expectation of a small healthcare network in a rural area differs from that of a big academic center. And as I discussed before in this chapter, expectations drive and, in many cases, affect satisfaction. In a related matter, the privacy of the doctor-patient relationship is now exposed, given the externalization of the doctor's role to be part of a group and health care system. The patient should be not only satisfied with his doctor but rather with the institution. The relationship is no longer confined to the doctor and the patient. I do remember being asked one day not to round on my patients without having the nurse with me because some studies showed that visiting a patient with the nurse increased patient satisfaction rates. On that day, I did think about the changes in the physician–patient relationship. I have no problem discussing the treatment plan with all the team members, including nurses, case managers, and pharmacists; frankly, I do that all the time. I also do believe that these multidisciplinary rounds are quite helpful in delivering better care, yet you can see the point I am trying to make. The relationship is no longer a two-party relationship.

Is there a lesson we can learn from these comparisons?

Despite the differences in the manifestations of this relationship, I believe it is still the core of medicine. A physician should prioritize the care of her patient over any other secondary gains. What protects this relationship and guides medicine is the doctor's character, knowledge, and virtues, which will always determine how she acts and direct her toward making an ethical decision. Medicine will always be a moral enterprise, whether it is delivered by a CMR machine of giant organizations or by a small leather bag in the streets of Aleppo.

Notes

1. Earl E. Shelp, The Clinical Encounter, The moral fabric of the patient—Physician relationship 1983, p. 3.
2. Pellegrino ED. Toward a reconstruction of medical morality. Am J Bioeth 2006;6(2):65.
3. Kaba R, Sooriakumaran P (2007) The evolution of the doctor patient relationship. International Journal of Surgery 5, pp. 58.
4. Earl E. Shelp, The Clinical Encounter, The moral fabric of the patient—Physician relationship 1983, p. 5.
5. Earl E. Shelp, The Clinical Encounter, The moral fabric of the patient—Physician relationship 1983, p. 7.
6. Earl E. Shelp, The Clinical Encounter, The moral fabric of the patient—Physician relationship 1983, p. 11.
7. Earl E. Shelp, The Clinical Encounter, The moral fabric of the patient—Physician relationship 1983, pp. 26–27.
8. Earl E. Shelp, The Clinical Encounter, The moral fabric of the patient—Physician relationship 1983, p. 15.
9. Earl E. Shelp, The Clinical Encounter, The moral fabric of the patient—Physician relationship 1983, p. 28.
10. Earl E. Shelp, The Clinical Encounter, The moral fabric of the patient—Physician relationship 1983, p. 18.

11. Earl E. Shelp, The Clinical Encounter, The moral fabric of the patient—Physician relationship 1983, p. 31.
12. Pellegrino ED, Thomasma DC. The conflict between autonomy and beneficence in medical ethics: proposal for a resolution. J Contemp Health Law Policy. 1987;3: p. 26.
13. Osorio, J. (2011). Evolution and changes in the physician-patient relationship. Colombia Médica, 42(3), 403.
14. Kaba R, Sooriakumaran P. The evolution of the doctor-patient relationship. Int J Surg. 2007 Feb;5(1):59.
15. Pellegrino ED, Thomasma DC. The conflict between autonomy and beneficence in medical ethics: proposal for a resolution. J Contemp Health Law Policy. 1987;3: p. 23.
16. Pellegrino ED, Thomasma DC. The conflict between autonomy and beneficence in medical ethics: proposal for a resolution. J Contemp Health Law Policy. 1987;3: p. 24.
17. Abraham Lurie. Consumerism in Medicine: Challenging Physician Authority Marie Haug Bebe Lavin. *Social Service Review. 59(1):150.*
18. Bodenheimer, T.: «The Industrial Revolution in Health Care» en Social Justice vol. 22 N° 4, 1995, pp. 26–33.
19. Cassell, E. (2004). The nature of suffering and the goals of medicine, second edition. P 77.
20. Marcum, J.A. 2008. An introductory philosophy of medicine: Humanizing modern medicine. New York: Springer, p. 65.
21. Lagay F. The legacy of humoral medicine. Virtual Mentor 2002.
22. Marcum, J.A. 2008. An introductory philosophy of medicine: Humanizing modern medicine. New York: Springer, pp. 66–68.
23. Boorse, C. (1997) A rebuttal on health. In: What is Disease?, Humber, J.M., Almeder, R.F., eds. Totowa, NJ: Humana Press, p. 4.
24. Sulmasy, D. 2005. Diseases and natural kinds. Theoretical Medicine and Bioethics 26: 489–498.
25. NESSE, R. M. & WILLIAMS, G. C. (1998). Evolution and the origins of disease. Scientific American 279, 86–93.
26. Marcum, J.A. 2008. An introductory philosophy of medicine: Humanizing modern medicine. New York: Springer, pp. 69–70.
27. Marcum, J.A. 2008. An introductory philosophy of medicine: Humanizing modern medicine. New York: Springer, pp. 63–64.
28. Cassell, E. (2004). The nature of suffering and the goals of medicine, second edition. P. 5.
29. Cassell, E. (2004). The nature of suffering and the goals of medicine, second edition. P. 81.
30. Cassell, E. (2004). The nature of suffering and the goals of medicine, second edition. P. 8.
31. Cassell, E. (2004). The nature of suffering and the goals of medicine, second edition. P. 35–36.
32. Emanuel, E. J. and Emanuel, L. L. (19923 Four models of the physician-patient relationship. Journal of the American Medical Association 267, 2221.
33. Marcum, J.A. 2008. An introductory philosophy of medicine: Humanizing modern medicine. New York: Springer, p. 281.
34. Marcum, J.A. 2008. An introductory philosophy of medicine: Humanizing modern medicine. New York: Springer, p. 282.
35. Pellegrino ED. Toward a reconstruction of medical morality. Am J Bioeth 2006;6(2):66.
36. Mallia, Pierre. 2012. The nature of the doctor—patient relationship, health care principles through the phenomenology of relationships with patients. P. 66.
37. Cassell, E. (1982). The nature of suffering and the goals of medicine. The New England Journal of Medicine, pp. 35–36.
38. Mallia, Pierre. 2012. The nature of the doctor—patient relationship, health care principles through the phenomenology of relationships with patients. P. 69.
39. Pellegrino ED. Toward a reconstruction of medical morality. Am J Bioeth 2006;6(2):67.
40. Mallia, Pierre. 2012. The nature of the doctor—patient relationship, health care principles through the phenomenology of relationships with patients. P. 32.

41. Mallia, Pierre. 2012. The nature of the doctor—patient relationship, health care principles through the phenomenology of relationships with patients. P. 33–35.
42. Mallia, Pierre. 2012. The nature of the doctor—patient relationship, health care principles through the phenomenology of relationships with patients. P. 37–41.
43. Mallia, Pierre. 2012. The nature of the doctor—patient relationship, health care principles through the phenomenology of relationships with patients. P. 43–46.
44. Mallia, Pierre. 2012. The nature of the doctor—patient relationship, health care principles through the phenomenology of relationships with patients. P. 49.
45. Pellegrino E, Thomasma DC. For the Patient's Good: The Restoration of Beneficence in Health Care. New York, NY: Oxford University Press; 1988. P. 101–102.
46. Marcum, J.A. 2008. An introductory philosophy of medicine: Humanizing modern medicine. New York: Springer, p. 289.
47. Reeder, Leo G. 1972. "The Patient-Client as a Consumer: Some Observations on the Changing Professional Client Relationship." Journal of Health and Social Behavior 13:407–408.
48. Marcum, J.A. 2008. An introductory philosophy of medicine: Humanizing modern medicine. New York: Springer, p. 290.
49. Pellegrino E, Thomasma DC. For the Patient's Good: The Restoration of Beneficence in Health Care. New York, NY: Oxford University Press; 1988. P. 51.
50. Veatch, R. M. (1972) Models for ethical medicine in a revolutionary age. Hastings Center Report 2, 5.
51. May, William F. *The Physician's Covenant: Images of the Healer in Medical Ethics.* 2nd ed. ed. Lexington, KY: Westminster John Knox Press, 2000. P. 125.
52. Marcum, J.A. 2008. An introductory philosophy of medicine: Humanizing modern medicine. New York: Springer, pp. 287–288.
53. Sitzia J, Wood N. Patient satisfaction: a review of issues and concepts. Soc Sci Med. 1997;45(12):1829.
54. Williams, S.J. and Calnan, M. (1991), "Convergence and divergence: assessing criteria of consumer satisfaction across general practice, dental and hospital care settings", Social Science Medicine, Vol. 33 No. 6, p. 707.
55. Goroll AH. Eliminating the term primary care "provider": consequences of language for the future of primary care. J Am Med Assoc. 2016;315(17):1833–1834.
56. Speeding. EJ, Rose DN. Buildingan effective doctor-patient relationship: from patient satisfaction to patient participation. Soc Sci Med 1985;21: 115.
57. Linder-Pelz, S. U. (1982a). Toward a theory of patient satisfaction. Soc. Sci. Med., 16, 577.
58. Bita Kash, Molly Mckahan. The evolution of measuring patient satisfaction. Journal of primary health care and general practice 2017, volume 1, p. 1.
59. Gill, L. and White, L. (2009), "A critical review of patient satisfaction", Leadership in Health Services, Vol. 22 No. 1, p. 8.
60. Ray Fitzparick, John Hinton, Stanton Newman, Graham Scambler, James Thompson. The Experience Of Illness. Oxford Tavistock, 1984, p. 155.
61. Williams B. Patient satisfaction: a valid concept? Soc Sci Med 1998;38:510.
62. Shelton P. Measuring and improving patient satisfaction. Gaithersburg, MD: Aspen; 2000, P. 3.
63. Stelfox HT, Gandhi TK, Orav EJ, Gustafson ML. The relation of patient satisfaction with complaints against physicians and malpractice lawsuits. Am J Med. 2005;118:1126–1133.
64. Shelton P. Measuring and improving patient satisfaction. Gaithersburg, MD: Aspen; 2000, P. 9.
65. Shelton P. Measuring and improving patient satisfaction. Gaithersburg, MD: Aspen; 2000, P. 2.
66. Reichheld, F.F. (1996). The loyalty effect: The hidden force behind growth, profits and lasting values. Boston: Harvard Business School Press. P. 13.
67. Shelton P. Measuring and improving patient satisfaction. Gaithersburg, MD: Aspen; 2000, P. 8.

68. LaVela SL and AS Gallan. Evaluation and measurement of patient experience. Patient Experience Journal. 2014; 1(1): 29.
69. Linder-Pelz, S. U. (1982a). Toward a theory of patient satisfaction. Soc. Sci. Med., 16, 578–579.
70. Sitzia J, Wood N. Patient satisfaction: a review of issues and concepts. Soc Sci Med. 1997;45(12):1834–1836.
71. Stimson, G and Webb. (1975) *Going to see the doctor: The Consultation Process in General Practice*. Routledge and Kegan Paul. London.
72. Sitzia J, Wood N. Patient satisfaction: a review of issues and concepts. Soc Sci Med. 1997;45(12):1834–1833.
73. Fitzpatrick R, Hopkins A. Problems in the conceptual framework of patient satisfaction research: an empirical exploration. Sociol Health Illness 1983;5:297–311.
74. Ben-Sira, Z. (1976). The function of the professional's affective behavior in client satisfaction: A revised approach to social interaction theory. Journal of Health and Social Behavior, 17, 3–11.
75. Gill, L. and White, L. (2009), "A critical review of patient satisfaction", Leadership in Health Services, Vol. 22 No. 1, pp. 9–10.
76. Ray Fitzparick, John Hinton, Stanton Newman, Graham Scambler, James Thompson. The Experience Of Illness. Oxford Tavistock, 1984, p. 160.
77. Gill, L. and White, L. (2009), "A critical review of patient satisfaction", Leadership in Health Services, Vol. 22 No. 1, p. 11.
78. Williams B. Patient satisfaction: a valid concept? Soc Sci Med 1998;38: 513.
79. Fitzpatrick R. Surveys of patient satisfaction: I – Important general considerations. BMJ 1991;302:887–889.
80. Sitzia J. How valid and reliable are patient satisfaction data? An analysis of 195 studies. Int J Qual Health Care 1999;11:319–28.
81. Sitzia J, Wood N. Response rate in patient satisfaction research: an analysis of 210 published studies. Int J Qual Health Care 1998;10:311–17.
82. Fenton J, Jerant A, Bertakis K, Franks P. The cost of satisfaction. Arch Intern Med 2012;172:405–411.
83. Fisher ES, Wennberg DE, Stukel TA, Gottlieb DJ, Lucas FL, Pinder EL. The implications of regional variations in Medicare spending. **Part 1**: the content, quality, and accessibility of care. Ann Intern Med. 2003; 138:273–287. Elliott S. Fisher, David E. Wennberg, Therese A. Stukel, Daniel J. Gottlieb, F.L. Lucas,, and Etoile L. Pinder. The implications of regional variations in medicare spending. **Part 2**: Health outcomes and satisfaction with care. Annals of Internal Medicine, 138(4):288–322, 2003.
84. Pellegrino E, Thomasma DC. For the Patient's Good: The Restoration of Beneficence in Health Care. New York, NY: Oxford University Press; 1988. P. 113.
85. Pellegrino E, Thomasma DC. For the Patient's Good: The Restoration of Beneficence in Health Care. New York, NY: Oxford University Press; 1988. P. 119.
86. Pellegrino, Edmund D., and David C. Thomasma. *Virtues in Medical Practice*. Cary: Oxford University Press, USA, 1993, pp. 4–5.
87. Pellegrino, Edmund D., and David C. Thomasma. *Virtues in Medical Practice*. Cary: Oxford University Press, USA, 1993, pp. 6–8.
88. Pellegrino E, Thomasma DC. For the Patient's Good: The Restoration of Beneficence in Health Care. New York, NY: Oxford University Press; 1988. P. 114.
89. Pellegrino, Edmund D., and David C. Thomasma. *Virtues in Medical Practice*. Cary: Oxford University Press, USA, 1993, pp. 9–13.
90. Pellegrino E, Thomasma DC. For the Patient's Good: The Restoration of Beneficence in Health Care. New York, NY: Oxford University Press; 1988. P. 64–65.
91. Pellegrino E, Thomasma DC. For the Patient's Good: The Restoration of Beneficence in Health Care. New York, NY: Oxford University Press; 1988. P. 66–68.
92. Pellegrino E, Thomasma DC. For the Patient's Good: The Restoration of Beneficence in Health Care. New York, NY: Oxford University Press; 1988. P. 78–82.

93.   Pellegrino E, Thomasma DC. For the Patient's Good: The Restoration of Beneficence in Health Care. New York, NY: Oxford University Press; 1988. P. 89–91.
94.   Pellegrino E, Thomasma DC. For the Patient's Good: The Restoration of Beneficence in Health Care. New York, NY: Oxford University Press; 1988. P. 88.
95.   Lewis, C. S. Mere Christianity: A Revised and Amplified Edition, with a New Introduction, of the Three Books, Broadcast Talks, Christian Behaviour, and Beyond Personality. 1st HarperCollins ed. ed. San Francisco: HarperSanFrancisco, 2001. P. 73.
96.   Pellegrino, Edmund D., and David C. Thomasma. *Virtues in Medical Practice.* Cary: Oxford University Press, USA, 1993, pp. 65–77.
97.   Pellegrino, Edmund D., and David C. Thomasma. *Virtues in Medical Practice.* Cary: Oxford University Press, USA, 1993, pp. 79–83.
98.   Lewis, C. S. *The Screwtape Letters.* Fontana Books. [London]: Collins, 1965.
99.   Pellegrino, Edmund D., and David C. Thomasma. *Virtues in Medical Practice.* Cary: Oxford University Press, USA, 1993, pp. 109–115.
100.   Pellegrino, Edmund D., and David C. Thomasma. *Virtues in Medical Practice.* Cary: Oxford University Press, USA, 1993, pp. 85–91.
101.   Pellegrino, Edmund D. (1999) The commodification of medical and health care: The moral consequences of a paradigm shift from a professional to a market ethic, Journal of Medicine and Philosophy, 24(3):243–246.
102.   Churchill, L. R. 1999. The United States Health Care System under Managed Care. How the Commodification of Health Care Distorts Ethics and Threatens Equity. Health Care Analysis Vol. 7 397.
103.   Pellegrino, Edmund D. (1999) The commodification of medical and health care: The moral consequences of a paradigm shift from a professional to a market ethic, Journal of Medicine and Philosophy, 24(3):248–255.
104.   Pellegrino, Edmund D. (1999) The commodification of medical and health care: The moral consequences of a paradigm shift from a professional to a market ethic, Journal of Medicine and Philosophy, 24(3):257–262.

Bibliography

1. Abraham, Lurie. 1985. Consumerism in medicine: Challenging physician authority Marie Haug Bebe Lavin. *Social Service Review* 59(1): 150.
2. Ben-Sira, Z. 1976. The function of the professional's affective behavior in client satisfaction: A revised approach to social interaction theory. *Journal of Health and Social Behavior* 17: 3–11.
3. Kash, Bita, and Molly Mckahan. 2017. The evolution of measuring patient satisfaction. *Journal of Primary Health Care and General Practice* 1: 1.
4. Bodenheimer, T. 1995. *The Industrial Revolution in Health Care* en Social Justice, vol. 22 N° 4, pp. 26–33.
5. Boorse, C. A. 1997. Rebuttal on health. In *What is disease?*, ed. J. M. Humber, and R.F. Almeder. Totowa, NJ: Humana Press, pp. 1–134.
6. Churchill, L. R. 1999. The United States health care system under managed care. How the commodification of health care distorts ethics and threatens equity. *Health Care Analysis* 7: 397
7. Emanuel, E. J. and L. L. Emanuel. 1992. Four models of the physician-patient relationship. *Journal of the American Medical Association* 267.
8. Fisher, Elliott S., David E. Wennberg, Therese A. Stukel, Daniel J. Gottlieb, F. L. Lucas, and Etoile L. Pinder. 2003. The implications of regional variations in medicare spending. Part 2: Health outcomes and satisfaction with care. *Annals of Internal Medicine* 138(4): 288–322.
9. Fenton, J., A. Jerant, K. Bertakis, and P. Franks. 2012. The cost of satisfaction. *Archives of Internal Medicine* 172: 405–411.

10. Fisher, E. S., D. E. Wennberg, T. A. Stukel, D. J. Gottlieb, F. L. Lucas, and E. L. Pinder. 2003. The implications of regional variations in Medicare spending. Part 1: The content, quality, and accessibility of care. *Annals of Internal Medicine* 138: 273–287.

11. Fitzpatrick, R., and A. Hopkins. 1983. Problems in the conceptual framework of patient satisfaction research: An empirical exploration. *Sociol Health Illness* 5: 297–311.

12. Fitzpatrick, R. 1991. Surveys of patient satisfaction: I—Important general considerations. *BMJ* 302: 887–889.

13. Gill, L., and L. White. 2009. A critical review of patient satisfaction. *Leadership in Health Services* 22(1).

14. Goroll, A.H. 2016. Eliminating the term primary care "provider": Consequences of language for the future of primary care. *Journal of the American Medical Association* 315 (17): 1833–1834.

15. Kaba, R., and P. Sooriakumaran. 2007. The evolution of the doctor-patient relationship. *International Journal of Surgery* 5 (1): 57–65.

16. Lagay, F. 2002. The legacy of humoral medicine. *Virtual Mentor.*

17. LaVela, S.L., and A.S. Gallan. 2014. Evaluation and measurement of patient experience. *Patient Experience Journal* 1 (1): 29.

18. Lewis, C. S., and Mere Christianity. 2001. *A revised and amplified edition, with a new introduction, of the three books, broadcast talks, Christian behaviour, and beyond personality.* 1st HarperCollins. San Francisco: HarperSanFrancisco.

19. Lewis, C. S. 1965. *The screwtape letters.* Fontana Books. [London]: Collins.

20. Linder-Pelz, S.U. 1982. Toward a theory of patient satisfaction. *Social Science and Medicine* 16: 577–582.

21. Mallia, Pierre. 2012. *The nature of the doctor—Patient relationship, health care principles through the phenomenology of relationships with patients.*

22. Marcum, J.A. 2008. *An introductory philosophy of medicine: Humanizing modern medicine,* 281. NewYork: Springer.

23. May, William F. 2000. *The physician's covenant: Images of the healer in medical ethics,* 2nd ed. Lexington, KY: Westminster John Knox Press.

24. Nesse, R.M., and G.C. Williams. 1998. Evolution and the origins of disease. *Scientific American* 279: 86–93.

25. Osorio, J. 2011. Evolution and changes in the physician-patient relationship. *Colombia Médica* 42 (3): 403.

26. Pellegrino, E., and D. C. Thomasma. 1988. *For the patient's good: The restoration of beneficence in health care.* New York, NY: Oxford University Press.

27. Pellegrino, E. D., and D. C. Thomasma. 1987. The conflict between autonomy and beneficence in medical ethics: Proposal for a resolution. *J Contemp Health Law Policy.*

28. Pellegrino, E.D. 2006. Toward a reconstruction of medical morality. *American Journal of Bioethics* 6 (2): 65.

29. Pellegrino, Edmund D., and David C. Thomasma. 1993. *Virtues in medical practice.* Cary, USA: Oxford University Press.

30. Fitzparick, Ray, John Hinton, Stanton Newman, Graham Scambler, and James Thompson. 1984. *The experience of illness.* Oxford Tavistock.

31. Reeder, Leo G. 1972. The patient-client as a consumer: Some observations on the changing professional client relationship. *Journal of Health and Social Behavior* 13: 407–408.

32. Reichheld, F.F. 1996. *The loyalty effect: The hidden force behind growth, profits and lasting values.* Boston: Harvard Business School Press.

33. Shelton, P. 2000. *Measuring and improving patient satisfaction.* Gaithersburg, MD: Aspen.

34. Sitzia, J., and N. Wood. 1997. Patient satisfaction: A review of issues and concepts. *Social Science and Medicine* 45(12).

35. Sitzia, J., and N. Wood. 1998. Response rate in patient satisfaction research: An analysis of 210 published studies. *International Journal for Quality in Health Care* 10: 311–317.

36. Sitzia, J. 1999. How valid and reliable are patient satisfaction data? An analysis of 195 studies. *International Journal for Quality in Health Care* 11: 319–328.

37. Speeding, E. J., and D. N. Rose. 1985. Buildingan effective doctor-patient relationship: From patient satisfaction to patient participation. *Social Science and Medicine* 21.
38. Stelfox, H.T., T.K. Gandhi, E.J. Orav, and M.L. Gustafson. 2005. The relation of patient satisfaction with complaints against physicians and malpractice lawsuits. *American Journal of Medicine* 118: 1126–1133.
39. Stimson, G., and Webb. 1975. *Going to see the doctor: The consultation process in general practice.* London: Routledge and Kegan Paul.
40. Sulmasy, D. 2005. Diseases and natural kinds. *Theoretical Medicine and Bioethics* 26: 489–498.
41. Veatch, R. M. 1972. *Models for ethical medicine in a revolutionary age.* Hastings Center Report 2.
42. Williams, B. 1998. Patient satisfaction: A valid concept? *Social Science and Medicine* 38: 509–516.
43. Williams, S. J., and M. Calnan. 1991. Convergence and divergence: Assessing criteria of consumer satisfaction across general practice, dental and hospital care settings. *Social Science and Medicine* 33(6).

Chapter 2
Models of Medical Clinical Practice: A Comparative Discussion of Secular and Religious Bioethics

Questions
- What are the goals of medicine?
- Considering the rapid progression of scientific tools and investigations, is there still a role for the history of presenting illness, and if so, why?
- Can we understand our role as physicians without a grand scheme of transcendent reference?

2.1 Introduction

In clinical practice, when many of us get totally absorbed in the busy schedules of the daily routines, appreciating the complexity of what we do, is not what we usually think about. I mean, in a sense, who has the time to analyze how he processes a patient's story, puts down a differential, and decides a treatment plan. All of that is usually performed by behaving professionally with patients, families, and surrounding colleagues. This routine of practice, which frequently become a pattern, can sometimes be challenged by complicated cases. Such cases that require a pause and a reflection. The way we practice medicine, think, decide, argue, and empathize, are all the result of a collective and pristine philosophy of medicine. This philosophy defines what medicine is at its core, its tools, goals, and most importantly, its code of ethics as a profession. This chapter discusses the philosophy of medicine, the epistemic theories of how doctors reason and think, and the different strengths and weaknesses in each of those models of intellectualization. All of which is usually done within a spirit of a transcendent vocation. In this discussion, I chose a Christocentric ethos to illustrate why medicine as a profession can be transformed into a vocation. However, I believe that the themes shared here are valuable, generalizable, and applicable regardless of the faith and beliefs of the practicing healthcare professionals.

S. Toro, *Introduction to Clinical Ethics: Perspectives from a Physician Bioethicist*,
https://doi.org/10.1007/978-3-031-30804-8_2

27

2.2 Philosophy of Medicine

Does the field of philosophy of medicine exist as a unique domain? And if so, what does it study? How it will help a practicing physician in his daily career? Is the philosophy of medicine merely a branch of the philosophy of science? Or it has, in its essence, something more to offer. The first section of this chapter will review the main views about the philosophy of medicine discussing from one side its relationship with the philosophy of science and, on the other side, its interaction with other fields such as sociology, psychology, and ethics.

2.2.1 General Consideration

In 1981, Pellegrino and Thomasma suggested that the contemporary crisis in medical practice had mainly resulted from the lack of unifying definition regarding the philosophy of medicine. They proposed trying to define the philosophy of medicine from the characteristics unique to medicine itself.[1]

In reviewing the literature, someone can encounter three main views regarding the philosophy of medicine as a field of study: a negative, a broad, and a narrow.[2] Thompson and Upshur subscribed to the first view. They suggested that the philosophy of medicine is simply a branch of the philosophy of science, like the philosophy of biology and the philosophy of physics. It deals with the three main domains of the philosophy of science, namely, Epistemology (the theory of knowledge and how it is acquired), Metaphysics (which in medicine would mainly inquire about monism and holism), and Logic (methodologies, theories, and scientific reasoning). In addition, they added Ethics as a fourth domain, which had gained significant attention since the 1950s.[3] In a similar analysis, Jerome Schaffer suggested that the problem created by medical practice would be best addressed by philosophers of science, philosophers of mind, and moral philosophers. Hence, he argued that there is nothing left for the philosophy of medicine to exist as a unique field of study.[4] Caplan also subscribed to the same view; he defined the philosophy of medicine as the study of epistemological and methodological dimensions of medicine. Though, he argued that the center of the philosophy of medicine has to be epistemological rather than ethical.[5]

The second view is a much broader one. Engelhardt agreed that philosophy of medicine studies the epistemological, logical, and axiological issues related to medicine.[6] However, he also broadened the definition to encompass areas of inquiry generated uniquely by medical practice, such as, the concept of health and illness, human suffering, logic of diagnosis, prognosis, and clinical trials. In a commentary on this view, Pellegrino argued that the Engelhardt definition of philosophy of medicine dilutes the specificity of it as a unique field of study. He suggested the third and narrow definition of philosophy of medicine. Being not only the summation of fields of philosophy of science, biology, ethics, aesthetics in a disorganized vision but rather a teleologically structured field that incorporates those areas of knowledge for a defined goal, which is the healing of an individual patient.[7]

Pellegrino made a distinction between four different models: philosophy and medicine, philosophy in medicine, medical philosophy, and philosophy of medicine. In the first one, medicine and philosophy, the two fields enter a mutual dialogue as separate entities.[8] Pellegrino argued that this dialogue between philosophy and medicine had resulted from positive and negative reasons. The positives are the common topics of interest shared between philosophers and physicians, like life, death, health, and suffering. The negative reasons are the conflict of methodologies and logic that both fields utilize; while medicine is mostly empirical and experimental, philosophical thinking is primarily abstract and analytical.[9] Throughout different eras, this relationship oscillated between harmony and discord. For example, Galen was serious about his identity as a physician and a philosopher of God. On the contrary, Hippocratic authors disgraced philosophical speculations and focused on searching for empirical evidence in their practice. In the Christian era, the relationship between philosophy and medicine became quite harmonious as both fields were considered the same. After the scientific revolution, the central interest of philosophy in medical practice shifted to studying the logic of diagnosis, statistical inference, causality, and nosology, focusing mostly on the scientific aspect of medicine.[10] Later on, after the 1950s, as a result of major scientific advances and growing of machines, there was a renaissance in the relationship between the two fields. Philosophers and theologians were re-invited to engage and help in dealing with major ethical debates that emerged mainly in the area of end-of-life care.[11]

The second model is philosophy in medicine. It is primarily the application of philosophical tools and methodologies in studying medical problems.[12] For example, studying the logic of medical diagnosis, ontology, and epistemology of various diseases. Another example is Beauchamp and Childress's principle-based system in contemporary bioethics, which had a profound impact on the ethical practice of medicine.

The third model, according to Pellegrino, is medical philosophy. It refers to the different styles of practice that physicians learn and gain through years of accumulative experience. For example, some clinicians are detailed oriented, holistic oriented, big picture oriented, and many physicians rather vary in their diagnostic or therapeutic enthusiasm. It is the search for practical wisdom, or in the old Greek terminology, the art and techne in medical practice, which cannot be fully grasped and appreciated by empirical evidence.[13]

The last model, the philosophy of medicine, seeks to understand what medicine qua medicine is. It uses philosophical methodologies to elucidate problems internal to medicine that cannot be fully grasped by the tools of medicine. Pellegrino argued that the are two differences between the philosophy of medicine and the philosophy of science. The first is factual, where the main goal of science is finding truth; the main goal of medicine is healing. The teloses of the two fields are quite different. Similarly, the philosophy of medicine is not also a philosophy of literature. For example, humanities study suffering and illnesses but not for healing a patient.[14] While someone can argue that a microbiologist, for instance, uses his tools to find antibiotics to heal a patient, the concept of healing is broader than just the biological healing.[15] Healing to make the patient a whole again, biologically, psychologically, socially, and spiritually.[16] Healing not in the general sense but the healing of a unique individual. This moves us into the second difference in the philosophy of medicine, which is phenomenological. It is the healing encounter that is established in the physician–patient relationship. The fact of illness, the act of healing, and the promise to heal.[17] The relationship in which the physician acts for the good of the patient. The unique good in its all dimensions and not the restricted and confined biomedical good.[18]

2.2.2 Reductionism and Holism

After illustrating the different notions of defining the philosophy of medicine, let us now move to closely examine two fundamental worldviews about clinical practice: Reductionism and Holism.

Reductionism, to begin with, can be traced back to Pre-Socratic atomists who held a belief that what is real is made up of *atomon* (unbreakable units of matter) and *void* (which is the space that separates those units).[19] Many centuries later, this methodology of thinking, was re-adopted by Descartes, who argued that the best way to understand the natural world is through studying its components. He gave a famous example of a clockwork machine and how someone needs to separately examine its parts and reassemble them in order to understand how the machine functions.[20] Isaac Newton also followed this approach, he said: "Truth is ever to be found in simplicity, and not in the multiplicity and confusion of thing." Studying the phenomena of the world in an isolated, simplified, and well-controlled environment (Newtonian physics: examining the forces on a perfect sphere, in a vacuum, moving down a smooth plane). In relation to this point, it is worth mentioning that Newtonian physics was closely associated with determinism. The notion that a phenomenon will always follow a certain path or behave in a certain way under the influence of the same factors. Same cause, same effect.[21] This harmonious association between reductionism and determinism continued to prevail until the beginning of the twentieth century, when it was challenged by the unpredictability of quantum mechanics.[22]

Reductionism, as a philosophy, can be divided into three types. Ontological reductionism: the idea that matter (molecules and their interactions) is the only existential reality. Henceforth, every biological system can be reduced to atoms.[23] For example, a tree is made up of wood, which in turn can be reduced cellulose, hemicellulose, and lignin. Cellulose is a polysaccharide made up of hundreds of D-glucose units, which can be further reduced to carbon, hydrogen, and oxygen atoms. The same approach can be applied to studying any human body organ, tracing it down to the level of tissues, cells, DNA, nucleotides, and atoms. The second type of reductionism is Methodologic reductionism: the idea that in order to understand a biological system, someone has to examine it at the lowest possible level. Therefore, accumulating the knowledge in down–top fashion. Lastly, Epistemological reductionism is the concept that a scientific domain can be reduced to another domain. For example, explaining human physiology in terms of physics and chemistry.[24] Evidently, since its wide adoption in the seventeenth century, reductionism as a philosophy has led to tremendous successes in the field of science and medicine. In fact, most of the modern discoveries in the field of physiology, microbiology, and molecular biology were achieved by taking this approach. However, confining the study of natural phenomena only to this reductionist model was not free of problems.

Critics of reductionism voiced the following concerns. As a methodology, reductionism replaces a broad understanding of many things with a detailed understanding of a few. Sometimes, it reaches a point where the explanation gets completely dissociated from the phenomenon. (20) Take the following example: imagine that a cholera epidemic occurred in our city; someone can extensively study the bacteria at the DNA and molecular levels, trying to understand its behavior and binary fission, identifying some modifiable factors, hoping that the results might help to control the spread. But this approach, apparently, might not be very helpful in studying the causes of the epidemic spread in the first place. (23) The phenomenon needs to be examined at a higher level of analysis. The same reasoning can be applied in approaching many diseases, like depression. Explaining the disorder solely as a decrease in the levels of serotonin in the brain is insufficient and very myopic.[25] Depression obviously is a much more complex psychological disorder than a decreased serotonin. Moreover, on a strictly biological level, reductionism as an approach fails to explain emergent properties. Biological organisms gain many characteristics as a result of the interactions between their components and the surrounding environment.[26] The characteristics of proteins, for example, cannot be fully explained by studying amino acids. Surface tension can't be explained by solely studying water molecules.[27] Moreover, reductionism takes a simplified approach by examining the phenomenon in a controlled environment. For example, studying cancer cells in the lab, isolating bacterial cultures, and testing antibiotics. Yet, many successful medications in vitro do not work in vivo.[28] The behavior of an organism in the real world is complex and always influenced by multiple factors. Ivan Illich emphasized this point. Modern medicine failed because it isolated itself from the messiness and complexity of open society.[29] These limitations in the reductionist approach led many scientists to start looking for alternative epistemic theories. The rise of system biology and system medicine in the last two decades are some fruits of such attempts. System biology aims to understand how molecules, organelles, and organs function in a system of complex networks.[30] It takes a holistic, anti-reductionist approach to examine natural phenomena.[31] Like reductionism, this philosophy also has ancient antiquity. It goes back to the Aristotelian notion that "the whole is more than the sum of its parts." Holism as a model of thinking did not only manifest in applying a new scientific methodology in studying medicine but also affected the way we start analyzing our approach to patient care. To illustrate more the weaknesses and strengths of Reductionism and Holism, I will provide some examples of how each of these theories approaches diseases and medical interventions.

Stegenga put it very nicely. Let us imagine an alien landed on earth and wanted to understand how a toaster works. He can simply observe how humans use the toaster, or he can dismantle it and study its compartments. Both approaches will provide a specific sort of knowledge that the other will not.[32] Reductionism gave us tremendous insight into understanding hereditary and infectious diseases. We now know the etiology of Sickle cell anemia, Syphilis, Phenylketonuria, and many other pathologies. However, we also know that causality is not as simple as it seems. Most chronic diseases are caused by a complex network of factors that interact with each other and manifest somehow uniquely in each individual. Take type I diabetes as an example. We know contributory genes like HLA-DRB1 and HLA-DQB1, but we also know that environmental factors predispose us to diabetes, such as bovine insulin and gluten. Moreover, the complexity of the disorder is not confined only to the biological level.[33] Holism, as an approach to studying disease, focuses on understanding the patient as a person.[34] Learning about his suffering and how the illness had affected him in a large socio-economic system.[35]

The second area to examine in the light of these two philosophies is medical interventions (treatments). Again, each of these theories offers a different kind of approach to suffering patients. If you had streptococcus pneumonia and got treated with Ceftriaxone, you should probably be very thankful to those who put a lot of effort into understanding the pathogenicity of the bacteria and discovered cephalosporins. However, if someone is malnourished, living in poor conditions, actively smoking, and keep presenting back and forth to the hospital with recurrent pneumonias, antibiotics treatment might not be solely the right answer for him. A similar analysis can be applied to various illnesses like depression, obesity, coronary artery disease, and many others. It is interesting to point out that the rise of evidence-based medicine (EBM) is one of the manifestations of taking the reductionist approach in studying interventions. An attempt to isolate, control, and study certain interventions/treatments in a specific patient population. In a similar analysis, the rising interest in patient-centered care[36] (PCC) and complementary medicine (CAM) can be viewed as a countermovement against the previous one. For some critics, the strict focus on biomedical aspects had resulted in dehumanizing medicine and diminution of care and compassion.[37]

2.3 Scientific Reasoning

When I started my fourth year of medical school, I remember very well being challenged, and quite often, by the high degree of uncertainty in clinical practice. As I moved away from the bench sciences (Anatomy, Physiology, Biochemistry), where everything seemed logical and predictable, to studying diseases, signs, symptoms, and highly complicated clinical diagnosis and management, became quite difficult. Now, while I am writing these pages, after almost twelve years, I think I have a better explanation of why practicing clinical medicine brought up frequently the phrase, "but this doesn't make any sense!!!".

2.3.1 Rationality and Empiricism

Rationalism and Empiricism are the main two epistemic positions in obtaining medical knowledge. In a way, as I will demonstrate later, the two views, at their core, complement and do not contradict each other. Rationalists, like Plato and Descartes, believed that knowledge is a priori knowledge.[38] It precedes our sensory experiences and is mainly obtained by the analytic functioning of the mind. This sort of knowledge is universal, immune to the biases of the senses, and can be applied in all circumstances. From a classification standpoint, there are two types of rational knowledge. Intuitive, which is present at birth, independent of the senses, like mathematical knowledge. And Innate, which is constitutive, is present at birth but gets elicited and moves to our conscious by experience. The arguments derived from such type of thinking are *deductive*; if the premises are true, then the conclusion is always true.[39] The famous example of this kind of argument: all men are mortal, Socrates is a man; therefore, Socrates is mortal. On the other hand, empiricists, like Hume and Locke, believed that knowledge is *posteriori* knowledge. For Locke, the mind was a *tabula rasa*, an empty slate upon which knowledge is written and gained through experience. Unlike rationalists, empiricists use *inductive* arguments in which the premises refer to a probabilistic conclusion. For example, if all the ravens in a random sample of 10,000 ravens are black, someone can strongly say that all ravens are probably black. The conclusion is never certain but probabilistic.[40]

By reviewing medical history, some can trace this debate back to the Greek era. Thessalus and Drakan, the sons of Hippocrates, emphasized the role of rationing in clinical practice. They stated that "when observations fail, reason might surprise."[41] The theory of Humors was a good example of the attempt to use this logic in explaining different illnesses. On the contrary, empiricists rejected this conceptualization. For them, if theoretical reasoning were enough, then philosophers would have been the best physicians. This group of doctors focused in their practice on clinical experience and recognizing patterns. Pharmacology and surgery were their preferred fields.[42] Throughout history, and over the years, these two models of thinking continued to enhance our knowledge. In fact, what we already know about pathophysiology and biochemistry is, to the greatest extent, a direct result of the successes of rationality.[43] On the other hand, the rise of epidemiology and evidence-based medicine in the late twenty century can be seen as a counter-movement of empiricism, where there was more focus on the outcomes and observations over the mechanistic understanding of illnesses.[44] The following passages will provide some examples of the dialogue between empiricists and rationalists.

2.3.2 Clinical Applications

In an argument regarding the benefits of using CYP2C19 genotype testing for guiding antiplatelet therapy in CAD patients, Nissen and Shuldiner had different opinions. Nissen emphasized the importance of evaluating the outcomes in a well-designed RCT before widely implementing CYP2C19 genetic testing into practice.[45] Shuldiner, on the other hand, believed that doctors should depend on their clinical judgment and best available evidence because delaying the application of a sound investigation might prevent some patients from getting beneficial therapies.[46] In 2013 when this debate was published, no randomized controlled trials were available to provide the gold standard evidence for answering such a question. In a comment on this argument, Prasad provided three notions of why the second view, which mostly depends on sound reasoning, is problematic.[47] First, the lack of gold standard RCTs will lead to inconsistency among clinicians; while some doctors will favor a treatment, others will not. The variation in the guidelines on the prophylactic indications of IVC filter placement is a good example of such an issue.[48] Second, the absence of evidence might encourage different parties to take financial advantage in promoting various treatments and equipment. Prasad gave an example of SSRI— A billion-dollar industry—in treating depression despite the lack of well-designed RCTs. Lastly, medical reversals are well-documented and known facts.[49] In the last few decades, many treatments, which seemed plausible from a purely physiological standpoint, ended up being ineffective and sometimes harmful when tested in large randomized controlled trials. Similarly, many medications widely thought to be contraindicated ended up being unexpectedly very beneficial. For example, beta-blockers were for a long time contraindicated in CHF patients.[50] This belief was based on a sound physiological background. Beta-blockers are negative inotropes, and logically, they should not be used in failing cardiac muscle. Two landmark studies later challenged this presumption, MERIT-HF[51] and COPERNICUS,[52] both showed that beta-blockers reduced mortality in these patients. The underlying rationale was the pathological effect of the inappropriately elevated sympathetic tone on left ventricular function, which can be reduced by using betablockers. Another example is PCI treatment for patients with stable angina. The approach was commonly used until the COURAGE trial showed that PCI with optimal medical therapy did not reduce the risk of death, MI, and major cardiovascular events compared to medical therapy alone.[53]

Despite its success in changing our contemporary practice, evidence-based medicine as a movement was placed under substantial criticism. Sniderman believed that EBM failed because physicians are not experts in this field, and they don't have sufficient time to evaluate the complex body of published literature. As a result, guidelines got created, which are, for the most part, not supported by gold standard RCTs.[54] Also, the results of many RCTs are conflicting, and the strict selection criteria of many of them have led to results that cannot be generalized to different populations. Moreover, most of RCTs are very expensive to conduct and are supported by the pharmaceutical industry, which raises many questions about conflict of interest and validity of the results. As a direct result of this financial advantage, EBM has weakened independent university research.[55] Lastly, there are sociological implications of applying EBM. The focus of quantifiable outcomes had shifted the trials away from the orphan fields like (psychology and sociology), where the interventions and outcomes are more challenging to measure. Consequently, an attempt was directed toward enhancing individual responsibility compared to collective-societal responsibility, such as (changing social and financial conditions). Such interventions and outcomes are harder to quantify and are less financially encouraging to funders, compared, for example, to immune modulators and chemotherapies.[56]

So which model of thinking should prevail? In reality, taking any of the extreme positions on the spectrum is clearly wrong. As Lanier and Rajkumar put it, the two methods should be present in each physician, with a pendulum swinging back and forth.[57] They don't compete with each other but form a positive feedback loop. A sound theory is formulated based on observations, then the theory gets tested, and outcomes might support or refute our previous understanding.[58] More data doesn't always mean more knowledge. We need rationality to connect and bridge our observations into theories.[59] Similarly, an RCT has to be based on a sound scientific hypothesis.[60] A strict focus on outcomes might decrease our creativity and imagination. As Hjørland[59] referred to in his article, "improving candlelight at the expense of discovering electricity." On the other side, we learned from our experience and practice that causality in medicine is extremely complex, predictability of outcomes is hardly certain, and what we are collectively convinced about today might revolutionary be changed tomorrow.

2.4 Subjective Reasoning

Aside from the scientific methodology of studying and approaching medicine, a subjective and less generalizable element significantly shapes our practice of medicine. For many reasons, some of which I had referred to, and some of which I will be discussing lately, medicine is not solely a science. And teleologically, the goals of medicine are not merely discovering facts for its own sake. In this section, I will examine the role of stories in medical thinking and give examples of how narratives are essential to the practical life of any physician.

2.4.1 The Importance of Stories

Humans are self-interpreting animals. We do not take the world surrounding us for granted. We continuously ask questions and try to make sense of our reality by telling and retelling sorties.[61] We tell narratives to communicate with each other. Our stories define who we are, shape us, and influence who we become.[62] Our narratives entail our values, beliefs, what we love, and what we hate. They help us to come to meaning and overcome the chaos of existence.[63] We always tell stories about our own lives. Every one of us strives to become the protagonist of his own story.[64] As a group of friends or a small community, we also have our own stories that tell who we are. When the population involved in the stories expands, the narratives become the narratives of societies and larger communities, reaching out to the historical level of countries, nations, and civilizations.[65]

Medicine, in its kernel, is essentially narrative. Physicians listen to patients' histories and learn about the diseases affecting their lives. They try to make sense of their presentations by setting events chronologically and gathering data to reach a consensus. All this is suffused with emotions of anxiety, fear, and empathy shared between the doctor and the patient.[66] Like every story, each clinical encounter contains characters, a conflict, and a resolution of that conflict. The characters are usually the patient, maybe a family and relatives, and the physician or physicians in charge.[67] The conflict is the illness in its wide presentation, and healing is the final chapter and the hopeful telos of the story. The patient is usually the protagonist of the narrative. He is the main character presenting with a disease, leading him to suffer. His suffering is not just physical pain, but a mysterious manifestation of losing functionality.[68] Suffering is an existential change in the person's perception of himself and the surrounding world.[69] The doctor tries to change the events of the story and help the patient regaining the functional component. By learning more about a patient's life, clinicians can better understand the patient's response. Our previous experiences significantly influence our reactions to life challenges. Why do some patients silently endure diseases and suffering, while others move to blame and despair? Why does someone fight while another surrender to metastatic cancer? All these behaviors are largely a reaction and response to previous stories.[70] By telling and listening to someone else narrative, we learn that we are not alone.[71] This form of sharing stories is in itself therapeutic. Many patients these days post their stories on Twitter, YouTube, and other forms of social media in an attempt to connect. AA meetings and addiction meetings are also known examples of healing experiences through sharing narratives.

Physicians also bring their own stories to the clinical encounter. Their personal and professional life experiences shape who they have become. The way they approach a case, deliver news, explain illness, and empathize with patients. All are collective phenomena of their lives. The art and philosophy of practice, how to joke, talk, and value patients' needs.[72] Robertson and Clegg put this idea nicely by making an analogy between physicians and actors. Whether we like it or not, patients judge us not only by our medical knowledge but also by the way we dress and act.[73] The white coat is not just simply a uniform. It is a part of a character trait of what is expected from a physician. We are all influenced and possessed, to some extent, by our physician's persona.[74] Like actors who have spent a great time practicing, learning how to move their body, their facial expressions, and voice tones, physicians spend years and years in training. Long hours of fatigue, distracting pagers, sleepless nights, all these experiences shape the persons they become. If you are a physician reading these pages, pose a second and reflect on the first day you took H&P from a patient. How do you think he or she evaluated your skills? It is not merely your medical knowledge that mattered that day but also who you were.[75] It takes years for us to learn how to empathize with different patients and clinical scenarios. Many times, we just can't do it.[76] This empathy and care are essential in our judgments because patients might not open up to us if they do not find their place in our narrative. Understanding patients' narratives help us to empathize, be less judgmental and make people more comfortable in telling us more about their suffering.[77]

2.4.2 Narrative Medicine Movement

After publishing "The Flexner Report "in 1910, medical education in the united states adopted the biomedical model as a gold standard approach to train physicians. The physician-scientist character became the ideal. The main focus was on embracing scientific methodology in understanding and treating various illnesses.[78] Over the last 100 years, this model prevailed and led to tremendous successes and achievements in health care delivery. However, there was a cost to pay by following that path. As Osler warned, "teacher and student chased each other down the fascinating road of research, forgetful of those wider interests to which a hospital must minister."[79] A sort of behavior many of us frequently encounter on a daily basis. The ethos of medical practice was at stake. As a result, over the past few decades, new movements in medicine started to raise voices. The focus was on physician–patient relationship, suffering, pain, humanities, and compassionate care. Narrative medicine movement is one of the manifestations of these efforts.

Rita Charon defines narrative medicine as "Medicine practiced with the narrative competence to recognize, absorb, interpret, and moved by the stories of illness."[80] Since medical practice is embodied in a narrative context, physicians will learn to practice better medicine by becoming narratively competent. Unlike the scientific approach, which aims to understand the patterns, mechanisms, and general truths about populations, narrative medicine focuses on the individual patient and the particulars.[81] Since science alone cannot grasp the full dimensions of human suffering, narrative medicine borrows the tools from humanities to help physicians in their practice.[82] By studying poetry, novels, and literature, the gained skills will enhance patient- physician relationship. It will also help clinicians to become more empathic and capable of delivering better compassionate care.[83]

Charon uses a technique of reflective writing called *Parallel Chart*. Health care professionals meet with professional trainers on regular bases to discuss their essays. Doctors choose to write about some encounters with patients. The writings are usually creative, reflecting their thoughts, emotions, and understanding of patients' suffering. This method aims to enhance compassion and solidify the patient-physician relationship.[84] In a study at Columbia University, twelve 4th year medical students got enrolled in a one-month elective rotation in narrative medicine. Students described the experience as enlightening and refreshing. They reported developing skills in listening and empathizing with each other and their patients. Students also stated that this experience enhanced their ability to recognize suffering in their patients and not merely to approach them as a case of a specific disease.[85] In another study of 50 Italian physicians writing parallel charts. 65% of the physicians reported learning new aspects about their patients, 15% stated that they were able to identify their errors and limits in delivering care. 74% thought that the technique made them realize how they could improve themselves and deliver more empathic care.[86]

In their book, *Storytelling in Medicine*, Robertson, and Clegg tell us a story about a young single mother-patient who had severe refractory asthma. This patient had frequent exacerbations with any sort of emotional distress. Throughout the treatment course, she got most of the side effects of systemic steroids without much clinical improvement in her main disease. The young lady worked in house cleaning, trying to support her two young children. The repeated exposure to dust and chemical was definitely not helping her condition. Despite being a patient for many years, it took her a long time to open up to her physician. She disclosed a story about her first daughter. A newborn, who was taken from her, when she was 16, and she had never seen her since. The authors, sharing this story, tell us that her doctor couldn't do anything medically more to help her. However, he was able to better empathize with her. He learned and understood in a way that probably no investigation could reveal why any emotional stress in the life of this lady was able to trigger severe asthma exacerbation.[87]

Mehl-Madrona also tells us an insightful story about his encounter with Melissa, a Cree woman with poorly controlled diabetes who suffered from coronary artery disease and ended up needing CABG. Madrona shares how Messila's life had changed after consequent life stressors. She learned about her daughters being sexually abused by her father. An event subsequently created chaos in the relationship with her mother. After that, her husband had an affair that broke the trust in their relationship. Melissa gradually stopped working on making beautiful jewelry and started overeating. She gained a lot of weight and ended up with diabetes. Madrona continues to inform us how sessions in meditation and reflection helped Melisa open up and disclose her feelings of shame and guilt, which were integrally linked to her disease. She found diabetes in her memories of the Catholic Church. He told us how making the analogy between her life and the traditional story of a coyote stealing fire, helped Melissa to understand the importance of insulin treatment. How she needed help and support from her surrounding community to overcome these life challenges, like the help offered to the First Women character by the frogs and Cranes. Through multiple sessions of reflection, Melisa's life changed dramatically; she started to walk and exercise; she went back to work and selling pieces of jewelry, and her Hemoglobin A1c went back to a normal range.[88]

2.5   Christian Perspective

2.5.1   *From Profession to Vocation*

What does it mean to be a professional? At present, the word has lost its pristine meaning. It became a common reference to a person who practiced a particular career for a long time and gained or excelled in a particular set of skills. Footballers, dancers, salesmen, and many others are currently called professionals.[89] Yet the roots of this word, historically and etymologically, had a unique meaning. The word *Profession* is derived from Profiteri, which literally means to declare aloud.[90] It is a promise that entails moral responsibility. Perhaps, medicine as a profession, has the most extended standing history of being a practice embedded in a structured moral obligation. Medicine is a moral enterprise because it deals with human suffering.[91] That sort of experience which leads to an existential crack in the holiness of a human being. When a person becomes ill, his body loses its transparency. The body becomes an obsessive object of thoughts and worries, a barrier standing against someone's life fulfillment and flourishing.[92] When a person becomes ill, he faces four facts: Finitude, Vulnerability, Dissolution of his own personhood, and Disruption of his presence in his community.[93] One of the most compelling questions, and perhaps the first to be asked, when someone experiences an illness, is how serious it is? Am I going to die? A dreadful challenge with the possibility of death. No matter how and where technology takes us, finitude is a reality, and everyone will eventually die.

The second fact in the experience of illness is Vulnerability. When one of us presents to the hospital or a clinic asking for help, he has already recognized that things went out of control. He has admitted his lack of power to take care of the issue, and consequently put himself in someone else's hands, asking for help. This phenomenon of the healing encounter is at the heart of all medical practice. It demands a fidelity of trust that the physician will not exploit the weak state of that patient. The relationship, as Pellegrino stated is unequal (90). The physician stands in a position of strength, and he ought to use the knowledge and experience he got for the service of his patient. That is why the ethics of market and contractarian models between patients and physicians cannot be radically applied in medicine. Someone can easily differentiate the state of an ignorant client walking into an electronics store to buy a laptop. He is somehow vulnerable because of his lack of knowledge about the different prices and qualities of computers. The seller, on the other side, is in a totally different position. He might or might not be aiming primarily for his own client's good. Nevertheless, his main goal at the end of this encounter is profit.[94] To make the best possible economic transaction. Later on, I will readdress the importance of this point while discussing the moral meaning of medicine as a profession.

The third fact in the experience of illness is the break of the patient's personhood. It is the existential change that he has to face. The person's autonomy—as a free moral agent—is put at stake when he gets labeled as sick. The dignity of a lady after mastectomy, the image of a grandmother with incontinence, the shame following a diagnosis of AIDS, and many other sorts of various burdening sicknesses. At this stage, the patient becomes dependent on the others. He now needs help, even in the most basic and essential daily activities. However, the damage doesn't not only stop there; it goes further and deeper. The patient is not only broken inside but also worried about losing those he loved. This takes us to the fourth fact: the disruption of his sense of attachment to the functioning community. Now he becomes a burden on his family, friends, and everyone else surrounding him.

The duty of the healer is to restore this broken relationship.[95] Sulmasy discussed this concept of illness in his article on biopsychosocial and spiritual approach to patient's care.[96] According to him, healing is a practiced art to make a person whole again, to restore the broken relationship inside and outside the patient. Sickness is not simply an abnormal lab or an Xray nodule. Sickness is a subjective perception of a disturbed state of functionality. Many patients continue to feel that something is wrong despite the lack of any detectable abnormalities in their tests (92). Although not commonly discussed at present, illnesses have closely attached metaphysical meanings in medical history. Sickness was mainly a spiritual experience caused by sinning against God.[97] In the thirteenth century, for example, a priest always accompanied the physician in offering healing services. The conviction was that once the patient faithfully repented, the disease will be cured.[98]

If we examine the concept of healing from a Christian standpoint, we can fairly say that the central message of the gospel is the good news of reconciliation. Jesus restored the broken relationship by his incarnation, crucifixion, and resurrection. His miracles were mainly healing miracles. For example, before healing the paralytic, he restored his relationship with the divine (Mt 9:1-8). He also corrected the broken relationship in the community when he answered the disciples' question of why a man was born blind. Jesus said that neither his nor his parents' sins but so the works of God might be displayed in him (John 9:1-3).[99] There is also another important theological aspect to be emphasized here. The intrinsic dignity of a human being as a son or a daughter of God. As I discussed above, illness exposes the weaknesses of the sick. At some point, there is a risk that patients might feel a lack of worth and dignity. Our duty as healers is to keep reminding them that no matter how dependent they become, their intrinsic dignity will always be present as sons and daughters of God.[100]

Now, how do all these ideas change the approach of a Christian physician to his life? According to Pellegrino[101] Four facts make medicine a profession: the fact of illness, the vulnerability of the patient, the necessity of fidelity, and lastly, the social contract the binds a physician to his community. The last fact needs further illustration. Medical knowledge is not proprietary. Clinicians and other health care professionals do not own it. The physician has stewardship over this knowledge to use it for the service of the sick. Society allows medical student to rotate on ill patients, tolerates the possible mistakes with new researches, gives physician certain privileges and resources because health is essential for a community to function. Hence, medicine as a profession, has an internal core of morality. For a Christian physician, medicine can not only be a profession but a vocation. A way of life dedicated to service and agape. Jesus emphasized the importance of taking care of the sick (Mathew 25:40). He asked his disciples to cure and spread the good news (Mathew 10:1–Mark 16:18). In Christian theology, healing is a divine act, an encounter with Jesus himself, for every patient is an alter Christus.[102] The Christian physician doesn't only carry the responsibility to try to heal and to be faithful to his talents, but also to witness the gospel in his life. To be one of Jesus disciples in spreading the good news.

2.5.2 Agape in Medicine

How do all these ideas find their way into a practical life? How a Christian physician ought to respond to a patient's vulnerability? As Sulmasy put it, the simple response to vulnerability is love (100). The altruistic love; that is not only as a sentiment but also a will. This sort of love that a Christian is called to appreciate and live his life through. A life that imitates the gospel of Christ. As Fr. Romano Guardini said regarding the Sermon on the Mount, "what Jesus has revealed there was no mere ethical code but a whole new existence."[103] The main message of Christianity is agape (Luke 10:27). Christians love their neighbors because they are sons and daughters of God. Agape neither seeketh her own nor asks for a utilitarian gain.[104] We can examine the practicality of Charity through the spectrum of beneficence in medical care. The least form of beneficence is benevolence, a mere kindness, a non-harmful attitude toward the other. Clearly, this position is not acceptable in clinical care, as physicians have a moral duty to act for the good of their patients and not only to avoid harm. When a physician moves further on the spectrum of care, he might try to act for his patient's good with a mild degree of inconvenience. The further he moves toward self-effacement in his service, the more he becomes altruistic, a real professional in his career. At the end of this spectrum comes the heroic form of love, Agape, the sacrificial love in its highest forms that resembles the love given to us by Jesus Christ. In a sense, Agapistic love is non-sensical and goes beyond logic and rationality. It is only possible through God's grace. It is crystal clear that the contractarian model of physician–patient relationship doesn't fulfill the required altruism in offering clinical care. For a Christian, medicine is not simply a career to make a living, it is a call, a vocation, a mission on the road toward the salvation of self and others.[105] It is worth mentioning here that physicians also have personal responsibilities in their lives, for themselves, for their own health and wellbeing, for their families and communities. The virtue of self-effacement has to be balanced with personal needs and self-interests.[106]

Agape also influences our understanding of another important ethical principle in medical care, which is *Justice*. From a philosophical standpoint, justice can be classified into five different forms: commutative, distributive, social, retributive, and compensatory.[107] *Commutative* justice addresses fairness in personal relationships. The mutual respect of duties and rights between individuals to avoid coercion and exploitation. In clinical practice, this form of justice plays a major role in the relationship between physicians and patients. A physician ought to treat patients fairly for their own good, and he shall not exploit his patients' vulnerability. The second form of justice, *Distributive*, as the word implies, deals with the fair distribution of resources in the society. Resources that are not owned by a particular group and are necessary for any society to function, like safety, food, housing, and healthcare. Two major theories significantly influenced our understanding of distributive justice at the present time. John Rawl's theory favors an egalitarian approach to iniquities in society, favoring the least advantaged population. On the other side, Nozick, rejected this notion favoring preserving the inadequacies in society. Rawls argued that his position is what most people would prefer in odds of a natural lottery, whereas, for Nozick, the natural lottery may lead to unfair distribution, but this is not necessarily unjust.

The third form of justice is social justice. It is the practical way of applying distributive justice. It functions by establishing committees, political parties, organizations, and many other forms of structured social bodies that aim to lead society toward a particular vision of life. Compensatory justice aims to find opportunities and rights for those who were not fortunate in the past. For example, if specific populations or individuals were excluded from having access to health care, schools or education. Compensatory justice affirms the responsibility of a society to compensate those people and find opportunities for them. Lastly, Retributive justice, addresses punishing those who committed violations or crimes against the community.[108] In health care, the concept can be applied, for example, to IVDUs, non-compliant patients, and many others to whom an argument can be formulated to not offer medical resources and treatments.

But how does agape change our understanding of justice? Why should we offer care to those who are considered the most vulnerable people in our community? Christocentric ethics has long antiquity in addressing this issue. As Veatch put it, it is challenging to defend the idea of taking care of the poor and least fortunate people in the community on a purely secular ground.[109] The principle of vulnerability and the common capacity to suffer might address partially this tendency.[110] We ought to behave justly and benevolently toward the others because we want them to do the same for us. This idea is commonly known as the Golden rule. Yet Christianity elevates this act to a much higher and noble level. This is exactly this group of people that Jesus asks us to care of, the poor, the sick, the unfortunate, and the sufferers. As Christians, we have a responsibility to care of the most unfortunate because we are all brothers and sisters in Christ.[111] This agapistic love that transforms and shapes the whole concept of Justice in life. For Christians, justice is rooted in agape and not in knowledge as Plato stated.[112] The dimensions of agape are not apprehensible by logic but rather possible through the grace of God. If we examine the philosophical concepts of Justice in light of Agapistic-Christocentric ethics, we can argue that it is closer to Rawl's position compared to Nozick's. Although Christian vision doesn't ask for utilitarian benefit. Retributive justice, on the other side, is not quite compatible with Agape. Withholding treatments and resources from IVDU's and other vulnerable individuals cannot be clearly defended on a Christocentric ground (107). Christian health care professionals are invited to live their lives as a call from God, a ministry, and a vocation. As Pellegrino put it, "medicine becomes a vehicle which raises the level of beneficence to Agape."

2.6 Summary

Is medicine only a science? Can medicine be studied and grasped in the full sense by scientific methods? How should a physician act in the light of uncertainties? This chapter tried to give a brief account of a few possible answers to those questions. The telos of medicine and the phenomenology of the healing encounter are two unique facts that give medicine its identity as a profession. It is neither totally a science nor mainly an art; it is both. In every clinical encounter, physicians need tools from both fields. Rationalization, narrative analysis, algorithmic thinking, and many more tools, along with accumulated clinical experience, form a unique philosophy of care. For a Christian physician, the responsibility of acting for the good of the patient requires a higher degree of altruism and self-effacement. A talent that he is called to work on and improve with faith and agape. Perhaps, with God's grace, he might one day experience the transformation of his career from a profession to a vocation.

Thought Box (II)
In daily practice, we often find ourselves quite overwhelmed by the system of work. There isn't always a time to self-reflect on what we do. Frequently there is not even enough time to take detailed stories from our patients and to meticulously organize our treatment plans. In addition to all our clinical duties, the daily routine is loaded with various meetings and extra clinical responsibilities. In the middle of all that, it is quite challenging for a doctor to sit down and assess what he is doing, why he is doing what he is doing, what is the big picture, and what his career goals are.

As colleagues, we rarely ask each other these questions. We prefer to talk about new clinical trials, mitral regurgitation, cardiogenic shock, and many other topics. We choose to discuss Twitter arguments and replies over delving into fundamental ideas. We shy away from any uncomfortable conversations about what we do and why. Many of us are so overwhelmed by building academic and research careers to the extent that if you ask someone why he is in fellowship, he will tell you because I want to become a tricuspid valve specialist or pulmonary hypertensionsit.

Indeed I am not trying to mock anyone here. I highly respect scientific and academic endeavors in the medical community, and I have some of that drive myself. Yet I am always careful to keep the main target while aiming at those different goals. As physicians, we should always remind ourselves that the telos of the practice is to reduce suffering and the helping those in need. Publishing, lecturing, educating, leading, and all of those achievements and duties should be done in congruence with that main goal. Whether a health care professional subscribes to the school of rationality, evidence-based medicine, or holistic care is not the main obsession. Whether you care for a patient by providing screening or clipping a mitral valve is not the primary issue. Citations are important, but they are not the main goal of the career. Practicing within a well-defined philosophy of medicine is essential and pivotal. No matter what specialty and what kind of practice we do, every part of our job should be viewed through that lens. In a sense, what I am suggesting might sound a bit spiritual; however, can we talk about our sickness without asking and having a spiritual experience? The role of the shaman might have died a long time ago when we discovered penicillin, but the dread of sickness and the awe of healing still accompany us each time we wear our white coats.

Notes

1. Weatherall D. J. (2004). Philosophy for Medicine. Journal of the Royal Society of Medicine, 97(8), 403.
2. Pellegrino, Edmund D., H. Tristram Engelhardt, and Fabrice Jotterand. 2008. The Philosophy of Medicine Reborn: A Pellegrino Reader. Notre Dame Studies in Medical Ethics. Notre Dame, Ind.: University of Notre Dame Press. P 26.
3. Thompson, R. Paul, and Ross Upshur. Philosophy of Medicine: An Introduction. London: Routledge, Taylor & Francis Group, 2018. 2018. Accessed October 28, 2019. P 5.
4. Marcum, James A. An introductory philosophy of medicine: Humanizing modern medicine. Vol. 99. Springer Science & Business Media, 2008. P 1.
5. Pellegrino, Edmund D., H. Tristram Engelhardt, and Fabrice Jotterand. 2008. The Philosophy of Medicine Reborn: A Pellegrino Reader. Notre Dame Studies in Medical Ethics. Notre Dame, Ind.: University of Notre Dame Press. P 27.
6. Pellegrino, Edmund D., H. Tristram Engelhardt, and Fabrice Jotterand. 2008. The Philosophy of Medicine Reborn: A Pellegrino Reader. Notre Dame Studies in Medical Ethics. Notre Dame, Ind.: University of Notre Dame Press. P 28–29.
7. Marcum, James A. An introductory philosophy of medicine: Humanizing modern medicine. Vol. 99. Springer Science & Business Media, 2008. P 4.
8. Pellegrino, Edmund D., H. Tristram Engelhardt, and Fabrice Jotterand. 2008. The Philosophy of Medicine Reborn: A Pellegrino Reader. Notre Dame Studies in Medical Ethics. Notre Dame, Ind.: University of Notre Dame Press. P 30.
9. Pellegrino, Edmund D., H. Tristram Engelhardt, and Fabrice Jotterand. 2008. The Philosophy of Medicine Reborn: A Pellegrino Reader. Notre Dame Studies in Medical Ethics. Notre Dame, Ind.: University of Notre Dame Press. P 24–25.
10. Pellegrino, Edmund D., H. Tristram Engelhardt, and Fabrice Jotterand. 2008. The Philosophy of Medicine Reborn: A Pellegrino Reader. Notre Dame Studies in Medical Ethics. Notre Dame, Ind.: University of Notre Dame Press. P 32.
11. Jonsen, Albert R. 2003. The Birth of Bioethics. New York: Oxford University Press. P 11.
12. Marcum, James A. An introductory philosophy of medicine: Humanizing modern medicine. Vol. 99. Springer Science & Business Media, 2008. P 2.
13. Pellegrino, Edmund D., H. Tristram Engelhardt, and Fabrice Jotterand. 2008. The Philosophy of Medicine Reborn: A Pellegrino Reader. Notre Dame Studies in Medical Ethics. Notre Dame, Ind.: University of Notre Dame Press. P 33–35.
14. Pellegrino, Edmund D., H. Tristram Engelhardt, and Fabrice Jotterand. 2008. The Philosophy of Medicine Reborn: A Pellegrino Reader. Notre Dame Studies in Medical Ethics. Notre Dame, Ind.: University of Notre Dame Press. P 36–39.
15. Pellegrino, Edmund D., H. Tristram Engelhardt, and Fabrice Jotterand. 2008. The Philosophy of Medicine Reborn: A Pellegrino Reader. Notre Dame Studies in Medical Ethics. Notre Dame, Ind.: University of Notre Dame Press. P 40–43.
16. Sulmasy D.P. 2002. "A Biopsychosocial-Spiritual Model for the Care of Patients at the End of Life." The Gerontologist 42: 24–33.
17. Pellegrino, Edmund D. "Toward a reconstruction of medical morality." The American Journal of Bioethics 6, no. 2 (2006): 65–71.
18. Pellegrino, Edmund D., and David C. Thomasma. For the Patient's Good: The Restoration of Beneficence in Health Care. New York: Oxford University Press, 1988. P 80–83.
19. Curd, P. (2007). Presocratic philosophy. In E. N. Zalta (Ed.), The Stanford encyclopedia of philosophy. Retrieved from http://plato.stanford.edu/entries/presocratics/.
20. Beresford, Mark J. "Medical reductionism: lessons from the great philosophers." QJM: An International Journal of Medicine 103, no. 9 (2010): 721.
21. Mazzocchi, Fulvio. "Complexity in biology." EMBO reports 9, no. 1 (2008): 10.
22. Thompson, R. Paul, and Ross Upshur. Philosophy of Medicine: An Introduction. London: Routledge, Taylor & Francis Group, 2018. 2018. Accessed October 28, 2019. P 43.

23. Fang, Ferric C., and Arturo Casadevall. "Reductionistic and holistic science." (2011): 1401.
24. Mazzocchi, Fulvio. "Complexity and the reductionism–holism debate in systems biology." *Wiley Interdisciplinary Reviews: Systems Biology and Medicine* 4, no. 5 (2012): 414.
25. Stegenga, Jacob. *Care and Cure: An Introduction to Philosophy of Medicine.* University of Chicago Press, 2018. P 68.
26. Mazzocchi, Fulvio. "Complexity in biology." *EMBO reports* 9, no. 1 (2008): 11.
27. Fang, Ferric C., and Arturo Casadevall. "Reductionistic and holistic science." (2011): 1402.
28. Mazzocchi, Fulvio. "Complexity in biology." *EMBO reports* 9, no. 1 (2008): 12.
29. Pilgrim, David. "The aspiration for holism in the medical humanities: Some historical and philosophical sources of reflection." *Health* 20, no. 4 (2016): 436.
30. Federoff, Howard J., and Lawrence O. Gostin. "Evolving from reductionism to holism: is there a future for systems medicine?." *Jama* 302, no. 9 (2009): 994.
31. Mazzocchi, Fulvio. "Complexity and the reductionism–holism debate in systems biology." *Wiley Interdisciplinary Reviews: Systems Biology and Medicine* 4, no. 5 (2012): 416–417.
32. Stegenga, Jacob. *Care and Cure: An Introduction to Philosophy of Medicine.* University of Chicago Press, 2018. P 72.
33. Smyth, Deborah J., Vincent Plagnol, Neil M. Walker, Jason D. Cooper, Kate Downes, Jennie H.M. Yang, Joanna M.M. Howson et al. "Shared and distinct genetic variants in type 1 diabetes and celiac disease." *New England Journal of Medicine* 359, no. 26 (2008): 2767–2777.
34. Wilson, Bruce. "Metaphysics and medical education: taking holism seriously." *Journal of evaluation in clinical practice* 19, no. 3 (2013): 481.
35. Stegenga, Jacob. *Care and Cure: An Introduction to Philosophy of Medicine.* University of Chicago Press, 2018. P 71.
36. El-Alti, Leila, Lars Sandman, and Christian Munthe. "Person centered care and personalized medicine: irreconcilable opposites or potential companions?." *Health Care Analysis* 27, no. 1 (2019): 45–59.
37. Miles, Andrew. "On a medicine of the whole person: Away from scientistic reductionism and towards the embrace of the complex in clinical practice." *Journal of evaluation in clinical practice* 15, no. 6 (2009): 942–945.
38. Marcum, James A. An introductory philosophy of medicine: Humanizing modern medicine. Vol. 99. Springer Science & Business Media, 2008. P 100.
39. Markie P. Rationalism vs. Empiricism. In Zalta E.D. (Ed), Stanford Encyclopedia of Philosophy (fall 2008 edition). Stanford: The Metaphysics Research Lab, 2008.
40. Hawthorne, J., 2012. Inductive logic. In: Zalta, E.N. (Ed.), The Stanford Encyclopedia of Philosophy. Center for the Study of Language and Information, Stanford, CA http://plato.stanford.edu/archives/win2012/entries/logic-inductive/.
41. Newton, Warren. "Rationalism and empiricism in modern medicine." *Law and Contemp. Probs.* 64 (2001): 300.
42. Newton, Warren. "Rationalism and empiricism in modern medicine." *Law and Contemp. Probs.* 64 (2001): 301.
43. Newton, Warren. "Rationalism and empiricism in modern medicine." *Law and Contemp. Probs.* 64 (2001): 302.
44. Newton, Warren. "Rationalism and empiricism in modern medicine." *Law and Contemp. Probs.* 64 (2001): 304.
45. Nissen, Steven E. "Pharmacogenomics and clopidogrel: irrational exuberance?." *JAMA* 306, no. 24 (2011): 2727–2728.
46. Shuldiner A.R., Vesely M.R., Fisch A. CYP2C19 genotype and cardiovascular events [letter]. JAMA. 2012; 307(14):1482; author reply 1485.
47. Prasad, Vinay. "Why randomized controlled trials are needed to accept new practices: 2 medical worldviews." In *Mayo Clinic Proceedings*, vol. 88, no. 10, pp. 1046–1050. Elsevier, 2013.
48. DeYoung, Elliot, and Jeet Minocha. "Inferior vena cava filters: guidelines, best practice, and expanding indications." In *Seminars in interventional radiology*, vol. 33, no. 02, pp. 065–070. Thieme Medical Publishers, 2016.

49. Prasad, Vinay, Andrae Vandross, Caitlin Toomey, Michael Cheung, Jason Rho, Steven Quinn, Satish Jacob Chacko et al. "A decade of reversal: an analysis of 146 contradicted medical practices." In *Mayo Clinic Proceedings*, vol. 88, no. 8, pp. 790–798. Elsevier, 2013.
50. Funck-Brentano, Christian. "Beta-blockade in CHF: from contraindication to indication." *European Heart Journal Supplements* 8, no. suppl_C (2006): C19–C27.
51. Merit-HF Study Group. "Effect of metoprolol CR/XL in chronic heart failure: metoprolol CR/XL randomised intervention trial in-congestive heart failure (MERIT-HF)." *The Lancet* 353, no. 9169 (1999): 2001–2007.
52. Packer, Milton, Andrew J.S. Coats, Michael B. Fowler, Hugo A. Katus, Henry Krum, Paul Mohacsi, Jean L. Rouleau et al. "Effect of carvedilol on survival in severe chronic heart failure." *New England Journal of Medicine* 344, no. 22 (2001): 1651–1658.
53. Boden, William E., Robert A. O'Rourke, Koon K. Teo, Pamela M. Hartigan, David J. Maron, William J. Kostuk, Merril Knudtson et al. "Optimal medical therapy with or without PCI for stable coronary disease." *New England journal of medicine* 356, no. 15 (2007): 1503–1516.
54. Sniderman, Allan D., Kevin J. LaChapelle, Nikodem A. Rachon, and Curt D. Furberg. "The necessity for clinical reasoning in the era of evidence-based medicine." In *Mayo Clinic Proceedings*, vol. 88, no. 10, pp. 1108–1110. Elsevier, 2013.
55. Hjørland, Birger. "Evidence-based practice: An analysis based on the philosophy of science." *Journal of the American Society for Information Science and Technology* 62, no. 7 (2011): 1305.
56. Parker, M. "False dichotomies: EBM, clinical freedom, and the art of medicine." *Medical humanities* 31, no. 1 (2005): 23–24.
57. Lanier, William L., and S. Vincent Rajkumar. "Empiricism and rationalism in medicine: can 2 competing philosophies coexist to improve the quality of medical care?." In *Mayo Clinic Proceedings*, vol. 88, no. 10, p. 1042. Mayo Foundation for Medical Education and Research, 2013.
58. Marcum, James A. An introductory philosophy of medicine: Humanizing modern medicine. Vol. 99. Springer Science & Business Media, 2008. P 103.
59. Hjørland, Birger. "Evidence-based practice: An analysis based on the philosophy of science." *Journal of the American Society for Information Science and Technology* 62, no. 7 (2011): 1306.
60. Sehon, Scott R., and Donald E. Stanley. "A philosophical analysis of the evidence-based medicine debate." *BMC health services research* 3, no. 1 (2003): 14. P 7.
61. Brockmeier, Jens, and Hanna Meretoja. "Understanding narrative hermeneutics." *Storyworlds: A journal of narrative studies* 6, no. 2 (2014): 6.
62. Robertson, Colin, and Gareth Clegg, eds. *Storytelling in medicine: How narrative can improve practice*. CRC Press, 2016. P 15.
63. Robertson, Colin, and Gareth Clegg, eds. *Storytelling in medicine: How narrative can improve practice*. CRC Press, 2016. P 2–3.
64. Mehl-Madrona, Lewis. *Narrative medicine: The use of history and story in the healing process*. Simon and Schuster, 2007. P 8.
65. Mehl-Madrona, Lewis. *Narrative medicine: The use of history and story in the healing process*. Simon and Schuster, 2007. P 52.
66. Brockmeier, Jens, and Hanna Meretoja. "Understanding narrative hermeneutics." *Storyworlds: A journal of narrative studies* 6, no. 2 (2014): 20.
67. Robertson, Colin, and Gareth Clegg, eds. *Storytelling in medicine: How narrative can improve practice*. CRC Press, 2016. P 4.
68. Árnason, Vilhjalmur, and Stefán Hjörleifsson. "The person in a state of sickness: The doctor-patient relationship reconsidered." *Cambridge Quarterly of Healthcare Ethics* 25, no. 2 (2016): 209–218.
69. Egnew, Thomas R. "The art of medicine: seven skills that promote mastery." *Family practice management* 21, no. 4 (2014): 27.
70. Robertson, Colin, and Gareth Clegg, eds. *Storytelling in medicine: How narrative can improve practice*. CRC Press, 2016. P 55.

71. Robertson, Colin, and Gareth Clegg, eds. *Storytelling in medicine: How narrative can improve practice.* CRC Press, 2016. P 93.
72. Mambu, Joseph. "Restoring the art of medicine." *The American journal of medicine* 130, no. 12 (2017): 1340–1341.
73. Robertson, Colin, and Gareth Clegg, eds. *Storytelling in medicine: How narrative can improve practice.* CRC Press, 2016. P 101.
74. Robertson, Colin, and Gareth Clegg, eds. *Storytelling in medicine: How narrative can improve practice.* CRC Press, 2016. P 102.
75. Robertson, Colin, and Gareth Clegg, eds. *Storytelling in medicine: How narrative can improve practice.* CRC Press, 2016. P 104.
76. Robertson, Colin, and Gareth Clegg, eds. *Storytelling in medicine: How narrative can improve practice.* CRC Press, 2016. P 110.
77. Robertson, Colin, and Gareth Clegg, eds. *Storytelling in medicine: How narrative can improve practice.* CRC Press, 2016. P 42.
78. Cassell, Eric J. "On the Destructiveness of Scientism." *Hastings Center Report* 45, no. 1 (2015): 46–47.
79. Duffy, Thomas P. "The Flexner report—100 years later." *The Yale journal of biology and medicine* 84, no. 3 (2011): 269.
80. Vannatta, Seth, and Jerry Vannatta. "Functional realism: A defense of narrative medicine." *Journal of Medicine and Philosophy* 38, no. 1 (2013): P 34.
81. Charon, Rita. *The principles and practice of narrative medicine.* Oxford University Press, 2017. P 156.
82. Barber, Sarah, and Carlos J. Moreno-Leguizamon. "Can narrative medicine education contribute to the delivery of compassionate care? A review of the literature." *Medical Humanities* 43, no. 3 (2017): 199–203.
83. Vannatta, Seth, and Jerry Vannatta. "Functional realism: A defense of narrative medicine." *Journal of Medicine and Philosophy* 38, no. 1 (2013): P 35–36.
84. Zaharias, George. "Learning narrative-based medicine skills: Narrative-based medicine 3." *Canadian Family Physician* 64, no. 5 (2018): P 354.
85. Arntfield, Shannon L., Kristen Slesar, Jennifer Dickson, and Rita Charon. "Narrative medicine as a means of training medical students toward residency competencies." *Patient education and counseling* 91, no. 3 (2013): 280–286.
86. Banfi, Paolo, Antonietta Cappuccio, Maura E. Latella, Luigi Reale, Elisa Muscianisi, and Maria Giulia Marini. "Narrative medicine to improve the management and quality of life of patients with COPD: the first experience applying parallel chart in Italy." *International journal of chronic obstructive pulmonary disease* 13 (2018): 287.
87. Robertson, Colin, and Gareth Clegg, eds. *Storytelling in medicine: How narrative can improve practice.* CRC Press, 2016. P 63.
88. Mehl-Madrona, Lewis. *Narrative medicine: The use of history and story in the healing process.* Simon and Schuster, 2007. P 238–250.
89. Pellegrino, Edmund D., and David C. Thomasma. *Helping and healing: Religious commitment in health care.* Georgetown University Press, 1997. P 32–33.
90. Pellegrino, Edmund D. "Toward a reconstruction of medical morality." *The American Journal of Bioethics* 6, no. 2 (2006): 67.
91. Pellegrino, Edmund D., H. Tristram Engelhardt, and Fabrice Jotterand. 2008. The Philosophy of Medicine Reborn: A Pellegrino Reader. Notre Dame Studies in Medical Ethics. Notre Dame, Ind.: University of Notre Dame Press. P 350.
92. Pellegrino, Edmund D., and David C. Thomasma. *Helping and healing: Religious commitment in health care.* Georgetown University Press, 1997. P 131–132.
93. Pellegrino, Edmund D., and David C. Thomasma. *Helping and healing: Religious commitment in health care.* Georgetown University Press, 1997. P 45.
94. Pellegrino, Edmund D., and David C. Thomasma. *Helping and healing: Religious commitment in health care.* Georgetown University Press, 1997. P 135.

48   2 Models of Medical Clinical Practice: A Comparative Discussion …

95. Sulmasy, Daniel P. *A balm for Gilead: Meditations on spirituality and the healing arts*. Georgetown University Press, 2006. P 22.
96. Sulmasy, Daniel P. "A biopsychosocial-spiritual model for the care of patients at the end of life." *The gerontologist* 42, no. suppl_3 (2002): 24–33.
97. Pellegrino, Edmund D., and David C. Thomasma. *Helping and healing: Religious commitment in health care*. Georgetown University Press, 1997. P 41.
98. Pellegrino, Edmund D., and David C. Thomasma. *Helping and healing: Religious commitment in health care*. Georgetown University Press, 1997. P 41.
99. Sulmasy, Daniel P. *A balm for Gilead: Meditations on spirituality and the healing arts*. Georgetown University Press, 2006. P 23.
100. Sulmasy, Daniel P. *The rebirth of the clinic: An introduction to spirituality in health care*. Georgetown University Press, 2006. P 33–34.
101. Pellegrino, Edmund D., and David C. Thomasma. *Helping and healing: Religious commitment in health care*. Georgetown University Press, 1997. P 91–93.
102. Sulmasy, Daniel P. *The rebirth of the clinic: An introduction to spirituality in health care*. Georgetown University Press, 2006. P 24.
103. Pellegrino, Edmund D., H. Tristram Engelhardt, and Fabrice Jotterand. 2008. The Philosophy of Medicine Reborn: A Pellegrino Reader. Notre Dame Studies in Medical Ethics. Notre Dame, Ind.: University of Notre Dame Press. P 352–353.
104. Pellegrino, Edmund D., and David C. Thomasma. *Helping and healing: Religious commitment in health care*. Georgetown University Press, 1997. P 147–148.
105. Pellegrino, Edmund D., H. Tristram Engelhardt, and Fabrice Jotterand. 2008. The Philosophy of Medicine Reborn: A Pellegrino Reader. Notre Dame Studies in Medical Ethics. Notre Dame, Ind.: University of Notre Dame Press. P 363–364.
106. Pellegrino, Edmund D., and David C. Thomasma. *The virtues in medical practice*. Oxford University Press, 1993. P 146–147.
107. Pellegrino, Edmund D., and David C. Thomasma. *Helping and healing: Religious commitment in health care*. Georgetown University Press, 1997. P 139–142, 155–156.
108. Walen, Alec. "Retributive justice." (2014).
109. Pellegrino, Edmund D., and David C. Thomasma. *Helping and healing: Religious commitment in health care*. Georgetown University Press, 1997. P 40.
110. Pellegrino, Edmund D., and David C. Thomasma. *Helping and healing: Religious commitment in health care*. Georgetown University Press, 1997. P 54–55.
111. Pellegrino, Edmund D., H. Tristram Engelhardt, and Fabrice Jotterand. 2008. The Philosophy of Medicine Reborn: A Pellegrino Reader. Notre Dame Studies in Medical Ethics. Notre Dame, Ind.: University of Notre Dame Press. P 365–366.
112. Pellegrino, Edmund D., and David C. Thomasma. *Helping and healing: Religious commitment in health care*. Georgetown University Press, 1997. P 149.

Bibliography

1. Árnason, Vilhjalmur, and Stefán Hjörleifsson. 2016. The person in a state of sickness: The doctor-patient relationship reconsidered. *Cambridge Quarterly of Healthcare Ethics* 25 (2): 209–218.
2. Arntfield, Shannon L., Kristen Slesar, Jennifer Dickson, and Rita Charon. 2013. Narrative medicine as a means of training medical students toward residency competencies. *Patient Education and Counseling* 91 (3): 280–286.
3. Banfi, Paolo, Antonietta Cappuccio, Maura E. Latella, Luigi Reale, Elisa Muscianisi, and Maria Giulia Marini. 2018. Narrative medicine to improve the management and quality of life of patients with COPD: The first experience applying parallel chart in Italy. *International Journal of Chronic Obstructive Pulmonary Disease* 13: 287.

4. Barber, Sarah, and Carlos J. Moreno-Leguizamon. 2017. Can narrative medicine education contribute to the delivery of compassionate care? A review of the literature. *Medical Humanities* 43 (3): 199–203.
5. Beresford, Mark J. 2010. Medical reductionism: Lessons from the great philosophers. *QJM: An International Journal of Medicine* 103 (9): 721–724.
6. Boden, William E., Robert A. O'Rourke, Koon K. Teo, Pamela M. Hartigan, David J. Maron, William J. Kostuk, Merril Knudtson, et al. 2007. Optimal medical therapy with or without PCI for stable coronary disease. *New England Journal of Medicine* 356 (15): 1503–1516.
7. Brockmeier, Jens, and Hanna Meretoja. 2014. Understanding narrative hermeneutics. *Storyworlds: A Journal of Narrative Studies* 6 (2): 1–27.
8. Cassell, Eric J. 2015. On the destructiveness of scientism. *Hastings Center Report* 45 (1): 46–47.
9. Charon, Rita. 2017. *The principles and practice of narrative medicine.* Oxford: Oxford University Press.
10. Curd, P. 2007. Presocratic philosophy. In *The Stanford encyclopedia of philosophy*, ed. E.N. Zalta. Retrieved from http://plato.stanford.edu/entries/presocratics/
11. De Young, Elliot, and Jeet Minocha. 2016. Inferior vena cava filters: Guidelines, best practice, and expanding indications. In *Seminars in interventional radiology*, vol. 33, no. 02, 065–070. New York: Thieme Medical Publishers.
12. Duffy, Thomas P. 2011. The Flexner report—100 years later. *The Yale Journal of Biology and Medicine* 84 (3): 269.
13. Egnew, Thomas R. 2014. The art of medicine: Seven skills that promote mastery. *Family Practice Management* 21 (4): 25–30.
14. El-Alti, Leila, Lars Sandman, and Christian Munthe. 2019. Person centered care and personalized medicine: Irreconcilable opposites or potential companions? *Health Care Analysis* 27 (1): 45–59.
15. Fang, Ferric C., and Arturo Casadevall. 2011. Reductionistic and holistic science. *Infection and Immunity* 79 (4): 1401–1404.
16. Federoff, Howard J., and Lawrence O. Gostin. 2009. Evolving from reductionism to holism: Is there a future for systems medicine? *JAMA* 302 (9): 994–996.
17. Funck-Brentano, Christian. 2006. Beta-blockade in CHF: From contraindication to indication. *European Heart Journal Supplements* 8 (suppl_C): C19–C27.
18. Hawthorne, J. 2012. Inductive logic. In *The Stanford encyclopedia of philosophy*, ed. E.N. Zalta. Stanford, CA: Center for the Study of Language and Information. http://plato.stanford.edu/archives/win2012/entries/logic-inductive/
19. Hjørland, Birger. 2011. Evidence-based practice: An analysis based on the philosophy of science. *Journal of the American Society for Information Science and Technology* 62 (7): 1301–1310.
20. Hompson, R. Paul, and Ross Upshur. 2018. *Philosophy of medicine: An introduction.* London: Routledge, Taylor & Francis Group. Accessed 28 Oct 2019.
21. Lanier, William L., and S. Vincent Rajkumar. 2013. Empiricism and rationalism in medicine: Can 2 competing philosophies coexist to improve the quality of medical care? In *Mayo clinic proceedings*, vol. 88, no. 10, 1042. Mayo Foundation for Medical Education and Research.
22. Mambu, Joseph. 2017. Restoring the art of medicine. *The American Journal of Medicine* 130 (12): 1340–1341.
23. Marcum, James A. 2008. *An introductory philosophy of medicine: Humanizing modern medicine*, vol. 99. Berlin: Springer Science & Business Media.
24. Markie, P. 2008. Rationalism vs. empiricism. In *Stanford encyclopedia of philosophy*, fall 2008 ed, ed. E.D. Zalta. Stanford: The Metaphysics Research Lab.
25. Mazzocchi, Fulvio. 2012. Complexity and the reductionism–holism debate in systems biology. *Wiley Interdisciplinary Reviews: Systems Biology and Medicine* 4 (5): 413–427.
26. Mehl-Madrona, Lewis. 2007. *Narrative medicine: The use of history and story in the healing process.* New York: Simon and Schuster.

27. Merit-HF Study Group. 1999. Effect of metoprolol CR/XL in chronic heart failure: Metoprolol CR/XL randomised intervention trial in-congestive heart failure (MERIT-HF). *The Lancet* 353 (9169): 2001–2007.
28. Miles, Andrew. 2009. On a medicine of the whole person: Away from scientist reductionism and towards the embrace of the complex in clinical practice. *Journal of Evaluation in Clinical Practice* 15 (6): 942–945.
29. Newton, Warren. 2001. Rationalism and empiricism in modern medicine. *Law and Contemporary Problems* 64.
30. Packer, Milton, Andrew J.S. Coats, Michael B. Fowler, Hugo A. Katus, Henry Krum, Paul Mohacsi, Jean L. Rouleau, et al. 2001. Effect of carvedilol on survival in severe chronic heart failure. *New England Journal of Medicine* 344 (22): 1651–1658.
31. Parker, M. 2005. False dichotomies: EBM, clinical freedom, and the art of medicine. *Medical Humanities* 31 (1): 23–30.
32. Pellegrino, Edmund D., and David C. Thomasma. *For the patient's good: The restoration of beneficence in health care.* New York: Oxford University Press.
33. Pellegrino, Edmund D., H. Tristram Engelhardt, and Fabrice Jotterand. 2008. *The philosophy of medicine reborn: A Pellegrino reader.* Notre Dame studies in medical ethics. Notre Dame, IN: University of Notre Dame Press.
34. Pellegrino, Edmund D. 2006. Toward a reconstruction of medical morality. *The American Journal of Bioethics* 6 (2): 65–71.
35. Pellegrino, Edmund D., and David C. Thomasma. 1997. *Helping and healing: Religious commitment in health care.* Georgetown University Press.
36. Pilgrim, David. 2016. The aspiration for holism in the medical humanities: Some historical and philosophical sources of reflection. *Health* 20 (4): 430–444.
37. Prasad, Vinay, Andrae Vandross, Caitlin Toomey, Michael Cheung, Jason Rho, Steven Quinn, Satish Jacob Chacko, et al. 2013. A decade of reversal: An analysis of 146 contradicted medical practices. In *Mayo clinic proceedings*, vol. 88, no. 8, 790–798. Amsterdam: Elsevier.
38. Prasad, Vinay. 2013. Why randomized controlled trials are needed to accept new practices: 2 medical worldviews. In *Mayo clinic proceedings*, vol. 88, no. 10, 1046–1050. Amsterdam: Elsevier.
39. Robertson, Colin, and Gareth Clegg (eds.). 2016. *Storytelling in medicine: How narrative can improve practice.* Boca Raton: CRC Press.
40. Sehon, Scott R., and Donald E. Stanley. 2003. A philosophical analysis of the evidence-based medicine debate. *BMC Health Services Research* 3 (1): 14.
41. Shuldiner, A.R., M.R. Vesely, and A. Fisch. 2012. CYP2C19 genotype and cardiovascular events [letter]. *JAMA* 307 (14): 1482; author reply 1485.
42. Smyth, Deborah J., Vincent Plagnol, Neil M. Walker, Jason D. Cooper, Kate Downes, Jennie H.M. Yang, Joanna M.M. Howson, et al. 2008. Shared and distinct genetic variants in type 1 diabetes and celiac disease. *New England Journal of Medicine* 359 (26): 2767–2777.
43. Sniderman, Allan D., Kevin J. LaChapelle, Nikodem A. Rachon, and Curt D. Furberg. 2013. The necessity for clinical reasoning in the era of evidence-based medicine. In *Mayo clinic proceedings*, vol. 88, no. 10, 1108–1110. Amsterdam: Elsevier.
44. Stegenga, Jacob. 2018. *Care and cure: An introduction to philosophy of medicine.* Chicago: University of Chicago Press.
45. Sulmasy, D.P. 2002. A biopsychosocial-spiritual model for the care of patients at the end of life. *The Gerontologist* 42: 24–33.
46. Sulmasy, Daniel P. 2002. A biopsychosocial-spiritual model for the care of patients at the end of life. *The Gerontologist* 42 (suppl_3): 24–33.
47. Sulmasy, Daniel P. 2006. *A balm for Gilead: Meditations on spirituality and the healing arts.* Washington, D.C.: Georgetown University Press.
48. Sulmasy, Daniel P. 2006. *The rebirth of the clinic: An introduction to spirituality in health care.* Washington, D.C.: Georgetown University Press.
49. Vannatta, Seth, and Jerry Vannatta. 2013. Functional realism: A defense of narrative medicine. *Journal of Medicine and Philosophy* 38 (1): 32–49.

50. Walen, Alec. 2014. *Retributive justice.*
51. Weatherall, D. J. 2004. *Book of the month: Philosophy for medicine*, 403–405.
52. Wilson, Bruce. 2013. Metaphysics and medical education: Taking holism seriously. *Journal of Evaluation in Clinical Practice* 19 (3): 478–484.
53. Zaharias, George. 2018. Learning narrative-based medicine skills: Narrative-based medicine 3. *Canadian Family Physician* 64 (5): 352–356.

Part II
Research and Biotechnology

Chapter 3
Ethics of Clinical Research, Publications, and Conflict of Interest

Questions

- Have you ever thought about the differences between the ethics of clinical practice and the ethics of scientific research?
- What is the concept of Equipoise?
- What are the necessary elements in informed consent? And what is the relationship between informed consent and a patient capacity?
- Mention two examples of a conflict-of-interest situation and suggest possible solutions?

3.1 Introduction

Medical research, whether translational or clinical, plays a major role in our clinical practice at present. Almost certainly, every clinical professional has been involved or is currently working on some research project. The number of medical journals expanded enormously in the last few decades, and physicians are continuously working on improving health care delivery and finding answers to many challenging clinical questions. However, the story of research development has not always been as colorful, shiny, and promising as it seems at present. We had to go through many trials to learn more about our human worth, dignity, and limitations. Some of those experiences were quite unjust and painful. And others could even be considered atrocities.

This chapter focuses on the ethical aspect of clinical research. First, it discusses the moral tension between the role of a clinician and researcher. Following that, it sheds light on the concept of Equipoise, the issues with placebo-conducted clinical studies, and the value and meaning of informed consent. Finally, it examines the theoretical basis and the moral psychology behind the problem of the conflict of interest research.

S. Toro, *Introduction to Clinical Ethics: Perspectives from a Physician Bioethicist*, https://doi.org/10.1007/978-3-031-30804-8_3

3.2   Historical Review

In 1753, Dr. James Lind, the surgeon on the British Navy Vessel Salisbury, performed the first documented clinical trial in medical history.[1] Dr. Lind took 12 sailors suffering from Scurvy and divided them into six groups. Each (two men) received one of those five remedies: nutmeg, seawater, cider, vinegar, elixir of vitriol, or citrus (two oranges and one lemon). After six days of treatment, the results of this small trial were obvious in the two patients who ended up receiving citrus. Both had improvement in their bleeding gums, fatigue, and cutaneous spots. In fact, one of them got so well and was appointed as a nurse to help the rest of the sick sailors.[2] Ironically, despite this evidence; it took the British Navy about 50 years to adopt a citrus rich diet.[3]

I am not sure if Dr. Lind obtained informed consent from those sailors. I am also reasonably confident that no independent IRB committee reviewed the research protocol. It would have been interesting to know why he chose to involve only twelve patients and why, specifically those lucky twelve sailors. Why did he choose two oranges and one lemon?[4] What was the anticipated benefit of drinking half a pint of seawater a day? Had Dr. Lind considered treating his research subjects fairly by reviewing his experiment's risk and benefit ratio? Imagining myself assisting this senior navy doctor on this trial, perhaps this sounds a little bit too much to ask for at that time. However, as someone might see, things had changed quite significantly since this eighteen-century scurvy experiment. The conduct of clinical research developed gradually over time and became remarkably organized and proto-coled. The course of this progress is too extensive to be summarized here; therefore, I will mainly focus my discussion on two influential historical projects: The Nazi experiments and the Tuskegee trail. Those two events will pave the way for us to thoroughly examine the grounding ethical theories behind clinical research in this chapter.

3.2.1   Some Tragedies

During WWII, many Nazi physicians were involved in conducting tragic research experiments on innocent war prisoners. Dr. Sigmund Rascher, a second lieutenant in the SS at that time, received direct permission from Heinrich Himmler to conduct experiments on living prisoners.[5] One of these trials aimed to test the response of humans to high-altitudes. Himmler wrote, responding to Rascher's request, "I can advise you that the prisoners will of course gladly be made available for the high-altitude flight research."[6] Dr. Rascher introduced for the first time what was later known as the "terminal experiments," which will end on tested persons' death.

The decompression chambers were shipped to Dachau concentration camp, where the study was planned to happen. The experiment aimed to evaluate the survival needs of research subjects by simulating parachute descending. Prisoners were tested at conditions mimicking 40,000 feet with and without oxygen. Changes in the subjects' physiological responses were documented, and the experiments were usually followed by postmortem autopsies.[7] In a letter to Himmler, Rascher wrote that such experiments could not be conducted on monkeys because they respond differently to such sorts of environmental stressors. This experiment on living humans was not the only one conducted. Similar experiments of military aims were also performed, including submerging prisoners in freezing water and testing their response to various rewarming methods.[8] Other unfortunate people were exposed to X-ray sterilization, Typhus infection, and drinking seawater trials. All those atrocities were sadly justified based on utilitarian and racial hygiene grounds. After WWII and in response to those tragedies, the Nuremberg code (1946) and deceleration of Hlesinki (1947) were established to protect the rights of research subjects and to emphasize the principle of informed consent and risk-benefit assessments.[9]

Another notorious study to address here is the Tuskegee trial. In 1932 the United States Public Health Service (PHS) conducted a research experiment to study the natural progress of syphilis in African American (AA) Population. At that time, physicians widely believed that syphilis manifested less aggressively in African Americans compared to Caucasians. By involving these patients in the study, the PHS hoped to show evidence supporting the virulence of Syphilis in (AA) population. The aim was to try to persuade the southern governments and legislators to fund syphilis treatment programs for AAs. So, by withholding treatment from a group of patients, a larger population might eventually get some benefits. The other reason for conducting the trial was to compare the clinical manifestation of advanced syphilis in AAs to white patients (Oslo Study). As there was a common belief that white patients suffered more from neurological symptoms compared to Black patients, who mainly suffered from cardiovascular pathologies.[10]

The study involved approximately 600 African American patients. Public health service started advertising and sending fliers in 1932. Patients were not informed about having syphilis disease. Instead, they were told that they had what was commonly known as "bad blood.". Informed consent was not obtained. PHS went even further and made sure to track the study participants and prevent them from receiving treatments after they moved out of Alabama.[11] Even after 1940, when Penicillin became available, research subjects were not treated. During the recruitment time in WWII, the PHS intervened to prevent study participants from joining the armed forces and receiving Syphilis treatment.[12] Also, to facilitate autopsies, PHS offered to cover burial costs and expenditures for low-income families. The study stopped in 1974 when a lawsuit case was filed on behalf of the study subjects. After hearing testimonies, the United States Congress established the National Commission for the protection of Human Subjects.[13] The Commission issued many reports and recommendations, including setting up IRB committees as independent reviewers of clinical research conduct.[14] It also issued the Belmont Report, which emphasized three principles in biomedical research: respect for Autonomy of study participants (informed consent), Beneficence (Risk/Benefit ratio), and Justice (selection criteria). (13)

Now after discussing some history, let us move to examine the grounding ethical problem in clinical research: the dilemma of the dual role of the Physician-Scientist.[15] The main concern of a practicing physician should be his patient's wellbeing.

The physician should strive to prioritize that goal over all other distracting commitments. The patient should always be treated as an end and never as a mean to another end, whether that end is money, prestige, or scientific knowledge. In contrast, the role of a researcher is dramatically different. The goal of science is to seek knowledge and reveal facts. In research, patients or subjects are mainly a mean to another end. The end is finding a generalizable consensus that might be applied to benefit other people.[16] To put these ideas in philosophical terminologies; we are referring to two major theories in medical ethics: Deontology and Utilitarianism. In the next section, I will try to examine these two theories and provide some examples to make the topic easier to digest.

·

3.2.2 Physician or Investigator, Ethical Analysis

Now after discussing some history, let us move to examine the grounding ethical problem in clinical research: the dilemma of the dual role of the Physician-Scientist.[17] The main concern of a practicing physician should be his patient's wellbeing. The physician should strive to prioritize that goal over all other distracting commitments. The patient should always be treated as an end and never as a mean to another end, whether that end is money, prestige, or scientific knowledge. In contrast, the role of a researcher is very different. The goal of science research is to seek knowledge and reveal facts. In research, patients or subjects are mainly a mean to another end. The end is finding a generalizable consensus that might be applied to benefit other people.[18] To put these ideas in philosophical terminologies, we are referring to two major theories in medical ethics: Deontology and Utilitarianism.

Remember the time when you watched The Dark Knight Movie? Didn't you wonder whether Batman should kill the Joker and end his problems? However, as everyone knows, Batman, have a fundamental rule: Never to kill. However, isn't it justified to sacrifice one person's life to save many others? Well, if someone is a Deontologist like Batman, the answer is No. However, if a Utilitarian thought about the question, the answer might be yes.[19] I tried to give this hypothetical example to simplify the concept. Of course, in medical practice, things are by far more complex than that.

One of the most famous scholars in deontology school is Immanuel Kant. A German philosopher who lived in the eighteenth century (1724–1804) and wrote a robust work on ethics and morality. Kant thought that morality could be derived solely from reason.[20] Because we are rational beings, we will all reach the same moral principles if we use this capacity. Kant disagreed with the empiricist philosophers of his time, such as Hume, who thought that our desires determine our morality. For Kant, morality was good intrinsically and not because it has an instrumental value that leads to preferred consequences.[21] Pojman and Fieser gave a fine example regarding this point. If two soldiers decided to cross the enemy's lines to deliver a message to the allies' troops on the other side of the battlefield, one of them succeeded, and the other got captured; do we consider the second less moral than the first because he failed in his task?[22] The value of morality for Kant was in the goodwill of the person who acted in accordance with the moral law and not in the consequences of the action.[23]

Kant introduced the concept of the *Categorical Imperative (CI)*. It is *Categorical* because it is absolute and unconditional, applied to everyone because we are rational beings.[24] In contrast, the hypothetical *Imperatives* are the sort of actions we do to achieve another goal (If you want A, do B). For example, if you want to succeed, then you have to study. If you want money, you have to work.[25]

The second word, *Imperative*, refers to a command which we ought to obey but might not. Kant articulated the CI in three formulas. The first is the principle of the law of nature: "Act as though the maxim of your action were by your will to become a universal law of nature." This means that the maxim (the rule guiding the action), must be universalizable before considering it moral. Kant gives an example of why people should not lie. If someone made a promise knowing that he could not fulfill (paying debt on a specific day), then he thought about universalizing this action on a societal level (everyone can make a promise knowing that he cannot fulfill); that will lead to what Kant referred to as a *Contradiction*. No rational person would agree that everyone should lie when making a promise, as there would be no meaning left for the act of promising itself; it will lead to a contradiction.[26]

The second formula in the Categorical Imperative is the principle of ends: to treat every individual as an *End* and never as a *Mean* to achieve another end. As we discussed before, the traditional relationship between a physician and a patient is very much deontological in that sense. The patient should never be used as a mean to achieve another goal (21). For Kant, humans have an ultimate value and dignity apart from any valuer's opinion. Their value is unconditional. They should not be manipulated and used to achieve a goal but rather always be treated as ends because of this ultimate value, which is based on their rationality.[27] The third formula in the Categorical Imperative is the principle of Autonomy. Kant conceptualization of Autonomy differs from the contemporary understanding of the notion.[28] For Kant, a person acts autonomously only if his drive was the goodwill of acting according to the moral law. For example, if a businessman insured his employees only to avoid possible lawsuits, then he did not act autonomously (18). If a drive influenced the act of a person other than a morality that fulfills the categorical imperative, then that person, according to Kant, had acted in a Heteronomy rather than Autonomy.[29]

Now let us move to the other theory, Utilitarianism. In contrast to the deontological theories, consequentialism, including utilitarian theory, considers morality a posteriori knowledge. It is not the nature act itself that determines whether it is right or wrong but rather the consequences of the act. The roots of this theory go back to the Greek philosopher Epicurus (342–270 BCE), who believed that our judgment of right and wrong is mainly determined by our natural tendency to seek pleasure and avoid pain.[30] Two main British philosophers articulated the Utilitarian theory, Jeremy Bentham (1748–1832) and John Stuart Mill (1806–1873).

Utilitarian theory is based on the principle of *Utility*: "we ought always to produce the maximal balance of positive value over disvalue, or the last possible disvalue."[31]

Both Bentham and Mill were Hedonistic; they understood values in terms of pleasure and happiness. Mill went further to divide pleasures into higher pleasures (intellectual and deep relationships) compared to less valuable pleasures (mere visceral, like food and sex). For Mill, although seeking out visceral pleasures might lead to temporary happiness, on the long-term, humans continuously search for higher values (Eudemonistic Utilitarianism, derived from the Greek word *Eudemonia* which means human flourishing).[32]

The Utilitarian theory is further divided into two main types: Act Utilitarianism and Rule Utilitarianism. For a strict Act Utilitarian, the morality of an action tends to be determined solely based on the consequences it produces. So, if lying to an elderly patient about a cancer diagnosis would produce less burden than disclosing the disease, then the action is morally plausible. In contrast, for a rule Utilitarian, the action is not simply the act of lying in a specific situation but rather the rule of lying in general. A rule Utilitarian would think about what harm might happen in the long term if we justified lying to patients as a rule. This might erode the fidelity between patients and physicians and lead to many harmful consequences despite that minor good that could have been produced initially by the act of lying.[33] The difference between strict rule Utilitarianism and Deontology is how the rule is determined to be right and wrong. For a Deontologist, in our example, lying is wrong because of the nature of the act (Priori judgment). On the contrary, for a Utilitarian, the rule is wrong because of its consequences (Posteriori Judgement).[34]

3.3 Randomized Controlled Trials

After elaborating briefly on the central grounding philosophical theories in clinical research, let us move now to examine two unique ethical dilemmas in randomized controlled trials: the concept of Equipoise and the problem of Placebo. Although many ethical challenges can be identified in the conduct of RCTs, for example (publication biases, scientific validity, conflict of interest, etc.), these two topics are frequently debated in the medical community, and they deserve close attention.

3.3.1 The Concept of Equipoise

The principle of Equipoise has been examined extensively in clinical literature in an attempt to soften the ethical conflict in the physician-investigator character. As I referred before to in my discussion, there is a deontological—Utilitarian dilemma that faces all physicians involved in the conduct of clinical trials. From one side, a physician should prioritize the care and welfare of his patient. On the other side, he should strive to maintain the scientific integrity and validity of the conducted research, which could often compromise his patients' individual interests. To point to this problem, scholars started referring to the principle of Equipoise as a necessity faculty in conducting RCTs.[35]

Equipoise is a state in which there is uncertainty about the merits of treatment or investigation. This uncertainty permits the enrolment of subjects in a research trial in order to generate generalizable and reliable knowledge.[36] In theory, the aim at the end of the trial is to disrupt the state Equipoise and show sufficient evidence that treatment A is more effective the treatment B or Placebo.

In the early 1980s, the medical community had a conflicting consensus in about the role of ECMO in treating newborns with severe respiratory failure. As a result, two important clinical trials were conducted to clarify and answer this question. Both of these two trials used a unique methodology called adaptive randomization.[37] In this form of randomization, the enrollment of patients into the experiment tends to favor over time the treatment arm that shows better results. Bartlett and his colleagues studied the role of ECMO in a group of twelve newborns with respiratory failure, comparing it to conventional therapy. Eleventh of studied patients received ECMO and survived; one patient received conventional treatment and passed away.[38] Three years later, another RCT was conducted at Harvard University to further confirm the superiority of ECMO in this patient's population. Neonates with persisted pulmonary hypertension (PPHT) were assigned to either receive conventional therapy in neonatal ICU or receive ECMO in the multidisciplinary ICU.[39] Physicians at each side of the treatment arm were comfortable in managing the researched patients according to their set of expertise (physicians in multidisciplinary ICU were comfortable using ECMO, whereas physicians in neonatal ICU preferred conventional therapy).[40] As I will be discussing later, this created a state of *Theoretical Equipoise* from a physician standpoint as each side of the research study was delivering the therapy that was believed to be superior. In phase I of this trial, nine patients received ECMO, and all survived, ten patients received conventional treatment, six survived, and four died. In Phase II and based on the study protocol (randomization should stop after four deaths in any arm), another twenty patients were assigned to ECMO, one patient died, and 19 survived. Accordingly, the survival was 97% in ECMO group compared to 60% in conventional therapy ($P < 0.05$). What was also interesting about these two trials was that they both used randomized consent. Physicians only asked the parents of newborns assigned to the ECMO arm to provide informed consent. The rationale was that newborns assigned to conventional arm were receiving what was accepted at that time a standard of care (they were not technically research subjects) and informing their parents about the option of ECMO would only exacerbate their emotional anxiety and burden. On the contrary, informed consent was obtained from all the parents of newborns assigned to ECMO treatment (39 consents were obtained). (37). The use of adaptive randomization (play the winner) approach in these two trials illustrates an attempt to alleviate the conflict between physicians' commitments. Because the investigators were not quite confident about the efficacy of ECMO they used this methodology to decrease the number of subjects assigned to the possible inferior treatment.[41] Also, the use of randomized consent could be seen as an attempt to deliver more compassionate care.

Now, from a classification standpoint, the concept of Equipoise can be viewed under four main categories, individual, clinical, patient-centered, and within the frame of consequentialist ethics. In 1974, Charles Fried argued that a physician has to be in a state of Equipoise in order to ethically enroll his patients in a clinical trial.[42] This understanding of the concept was later termed *Individual* or *Theoretical* Equipoise.[43] Many scholars criticized this vision arguing that confining the judgment about equipoise to each individual physician is not practical. First, physicians usually have different opinions regarding various medications. Second, finding a 50/50 percent evidence supporting each arm of the trial is usually theoretical and needs a simple null hypothesis, which is almost always not the case. Furthermore, the individual-physician judgment about the studied treatment might change during the conduct of the trial; in that case, the physician has an ethical duty to ask his patients to voluntary withdraw from the study.[44]

Three years later, Benjamin Freedman proposed *Clinical Equipoise* as an alternative approach. For him, it is not the judgment of an individual clinician that determines whether the offered treatments are in a state of Equipoise but rather the overall consensus of the medical community.[45] If there is no general agreement between experts regarding which treatment is better, then it is ethically permissible for a physician to enroll his patients in a clinical trial. This approach to the problem, although definitely sounds more solid, was also later criticized. The argument was based on the fact that when a patient seeks help from a physician, he asks for his personal care and opinion regarding an illness and not merely the opinion of the medical community.[46] He voluntarily enters into a relationship guided by fidelity and puts his vulnerability and suffering in the hands of his doctor. Although this objection might not be too convincing, as there is always a common standard of care guiding physicians' practice, it again unmasks the internal tension in the role of physician-investigator.

Other scholars tried to shift the judgment toward the patients and their representatives, I will name this approach *Patient Centered*. For them, the research subjects have to determine whether or not the suggested treatments are in a state of Equipoise. Informed Consent has a major role in this argument.[47] Again, someone can see that this approach is also very fragile. Although patients' values should be considered in setting a therapeutic plan, confining the decision solely to a patient is not practical and even unethical. Patients need their physicians 'care, compassion, experience, and wisdom. They cannot make decisions on their own, and that is why they seek help.

Lastly, some scholars went differently in their attempt to justify the ethics of randomization. Marquis suggested adopting consequentialist ethics in defending randomization. Meier and Royal argued in favor of altruism, arguing that many patients will be willing to compromise and sacrifice to some extent their personal gains to enhance knowledge and improve care for future patients. Other scholars proposed Social Contract as a theory to justify the conduct of research. They argued that most of us would prefer to live in a society in which there is a continuous potential for improvement in treatments compared to a society that offers only what is known to be effective. (45)

Before closing this discussion, I would like to emphasize a point discussed by Freedman.[48] The practice of medicine is social rather than individual. In a sense, there are standards of care agreed upon by the professional medical community. Physicians earn their privileges to practice in society after they acquire a certain level of knowledge and pass required exams. They don't practice based on what they solely like and prefer. Most of the time, physicians have to sacrifice, to a modest degree, the care of some patients for the common good of the other patients. For example, not spending two hours with each patient in the clinic in order to see more patients and use time more judiciously.[49] Another example would be the prioritization of resources for the sickest patients in the hospital. What is of extreme importance in conducting RCTs is not to exploit the vulnerability of research subjects, respect their dignity and always try weighing out the potential risks and benefits before exposing patients to experimental treatments.[50]

3.3.2 The Problem of Placebo

In the current era of Evidence Based Medicine (EBM), the general focus of the medical community has been shifted toward obtaining robust scientific evidence regarding the effectiveness of various therapeutic remedies. However, rigorous evidence is not always easily achievable, and since its adoption as a gold standard in EBM, randomized controlled trials (RCTs), introduced a plentitude of ethical, methodological, and epistemological challenges. Indeed, one of those problems is the problem of Placebo.

Historically, the first reference to using placebo in medical practice goes back to the early eighteenth century. The term was used by Dr. Alexander Sutherland in criticizing some fashionable physicians of his age.[51] Around the same time, William Cullen, another English physician, referred to using Mustard Powder on one of his patients as placebo. In 1977, the first placebo-controlled trial was conducted by Dr. John Haygarth to debunk animal magnetism in treating boils.[52] In 1920, about a decade later, the reference to the possible psychological effect of various medications came into attention through the work of T. C. Graves. Gradually, as medicine became more scientific during the twentieth century, the reliance on RCTs increased substantially. Initially, when there was not robust evidence grounding many prescribed medications, the use of placebo in RCTs did not create a huge problem. However, later, when we started accumulating our medical knowledge and re-forming our medical ethics, conflicts around the topic began to arise.

Generally speaking, there are few conditions in which the conduct of placebo-controlled trials is not controversial: (1) when there is no established effective treatment, (2) when patients fails to respond to known treatments,[53] (3) when the studied disease is not serious enough to cause significant harm if known effective medications were withheld (baldness, allergic rhinitis),[54] (4) when the response to known medications is so variant that using an active-controlled therapy might not lead to scientifically valid results (In treating depression, for example, many well-designed trials failed to demonstrate the effectiveness of antidepressants compared to placebo). On the contrary, there are situations in which withholding known effective treatment from research participants is definitely not acceptable. For example, assigning patients post Myocardial infarction to Placebo (Not giving them Aspirin/Plavix) to test the efficacy of a new antiplatelet therapy.[55] However, the problem of placebo becomes an issue in those in-between situations, where there is no consensus about the effectiveness of a medication or when the evidence of a known therapy became questionable by a new discovery, warranting further investigation.

Proponents of Placebo defend their positions through many arguments. The most appealing one is the argument of Assay Sensitivity: which means the ability of a clinical trial to establish an absolute efficacy, a bench march, that the tested medication is effective.[56] This problem becomes quite evident in non-inferiority trials where external evidence (outside the conducted study) is required to confirm the findings of the conducted study (If X medication is non inferior to Y, that would mean nothing unless Y is known to be effective from the beginning). Some scholars rejected this argument claiming that the placebo effect is not consistent and can vary between trials, ergo the acquired evidence cannot be used as a benchmark. Also, they rejected the Duhamian argument of external evidence, claiming that even a placebo-controlled trial uses a significant body of assumptions not tested in the trial itself, ergo the external evidence argument is not as valid as it sounds.[57] Opponents of placebo-controlled trials also defend their positions by mainly referring to Clinical Equipoise and the necessity to avoid harming patients by withholding effective medications. Freedman argued that clinical research should be carried out to promote healing and not to discover mere biological facts.[58] Undoubtedly when an effective established treatment is available, it should not be withheld from research subjects. However, as I referred to earlier in this chapter, the problem becomes apparent when the evidence is not robust. In those circumstances, the argument for using placebo could have more weight if someone considered the possibility of obtaining a better statistical strength with lesser research subjects, thereby exposing a smaller number of patients to experimental medications.[59]

3.4 Informed Consent

3.4.1 Grounding Theories

Before examining the philosophical bases of Informed Consent, it is crucial to provide some historical facts. Until the beginning of the twentieth century, and even later on, the conduct of medicine was strictly paternalistic to some extent. The physician took full responsibility for his patient's care and solely decided what sort of therapy a patient should receive. Patients' values and beliefs were not quite important at those times. The American Medical Association code of ethics in 1847 insisted on the necessity of implicit obedience of patients to their doctor's prescriptions. A common postulation was that a patient should only be told what is good for him, which is primarily determined by his physicians.[60] Of course, things are pretty different at present. The principle of Autonomy plays a major role in medical practice and in determining treatment plans. The role of informed consent in clinical research is an essential requirement, both morally and legally. As I discussed before, in response to the atrocities of the Nazi experiments in WWII, the Nuremberg code (1946) and the deceleration of Helsinki (1947) were established to protect the rights of research subjects and to emphasize the principle of informed consent (9). So, someone can grasp that IC, in a sense, provides a sort of protection and enhancement of social justice by strengthening general trust in the practice of medicine. IC also has a major impact on individual care. When a patient communicates his values to his physician, the latter might be able to provide better care by making sure that this patient will receive the treatment that is congruent with his values. So, IC is essentially grounded on beneficence.[61] In addition, informed consent is also majorly grounded on the principle of Autonomy. We ought to respect the dignity of humans because they are rational subjects capable of reasoning and formulating values.[62] Humans have intrinsic value by the sole fact of being humans.[63] The patient has the right to determine what should be done to his body and what is the best treatment option for him. Of course, things are not as simple as we hope. Many factors affect the ability of a patient to give informed-knowledgeable consent, and many scholars criticized the concept and even questioned its overall validity. Before I move to the next section and discuss the various elements of informed consent, I would like to emphasize an important idea articulated by Edmund Pellegrino in his article *Toward a Reconstruction of Medical Morality*.[64] The etymology of the word Consent comes from the Latin words (Con-Sentire), which means to feel together. Pellegrino pointed out that informed consent should not only be understood as a legal or medical document containing certain information that meets specific criteria. Instead, it is a moral document reflecting the mutual trust that is formed between a patient and a physician when they enter into the healing relationship.

3.4.2 Essential Elements

What does it mean to give INFORMED consent? How much information should be provided for a patient to give valid consent? Does withholding some information considered a betrayal of mutual trust? As someone can see, all those questions are valid questions. To make this point clearer, let us take the following example. A patient went to the oncologist office to discuss getting involved in a clinical trial testing new Chemotherapy named XYZ. The doctor was extremely meticulous in providing all the details of the study. In fact, he reserved a three-hour appointment to discuss with the patient the treatment plan. The doctor went all the way in explaining the meaning of RCT, the methodologies, statistical analysis, all the details about how the Chemotherapy works, and the possible benefits and risks that the patient might experience. Can someone imagine the patient's situation after those three hours?

Would he really be informed? Of course, this is an exaggerated example. However, the main point here is that the content of the consent should be written to help a patient make decisions.[65] Not for the sake of only giving information. The abstract notion of knowledge is not to be discussed in that case, as discussing anything in life can be endless and be viewed in many different ways. The doctor's duty is to enhance his patient's autonomy by making the knowledge easier to digest.[66] Again, no matter how detailed IC gets, which is usually done for legal rather than medical purposes, there is a component of mutual trust that cannot be avoided.

The Second element in informed consent is Voluntariness: ensuring that a patient is not getting involved in a trial under coercion or pressure. Of course, it is quite important here to refer to patients' vulnerability, especially for those groups of patients with mental diseases, prisoners, children, or soldiers.[67] When a patient gets involved in a clinical trial, he enters the trial expecting some benefits. However, sometimes, his consent to enter the trial might be influenced by an external source other than the possible benefit of the trial itself. For example, would it be ethical to try to convince a prisoner to enter into a study as the only way of getting medical care or promising a homeless to get a significant amount of money if he enters a clinical trial? Freedman makes an important point regarding this point when he discusses the fundamental rights of each individual (Like basic health care, safety, freedom of choice, and religion). If those basic rights are not already met in a research subject, then he is at high risk of being influenced by any reward. Therefore, his physician should be extremely careful in examining such a patient drive to be enrolled in a clinical study.[68]

Lastly, any research participant should have the capacity to give his or her informed consent. A patient has capacity when he can reflect on his life choices, reason the possible benefits/risks and change his judgments when the situation changes.[69] Capacity is a complex concept. A given patient might have the capacity to provide consent for certain procedures and not others depending on the degree of procedural/treatment complexity. A patient might be able to carry on his basic life activities but lack the capacity to make more complex decisions. In clinical research, the more a trial involves potential risks, the more a physician should be careful in assessing patient's capacity. For example, involving a patient in a clinical trial of eczema—in terms of capacity assessment—would be totally different from involving the same patient in a clinical trial of new immunotherapy or new generation of cardiac stents.[70]

3.5 Conflicts

3.5.1 The Vioxx Story

In 1999, Merck pharmaceutical company developed Roficoxibe (Vioxx), a cyclooxygenase -2 (COX-2) inhibitor, as a safer option compared to NSAIDs (Non-steroidal Anti-inflammatory Agents) in treating patients with Osteoarthritis.[71] By selectively blocking COX-2, Vioxx helps to relieve pain and decreases joints inflammation without compromising the integrity of the gastrointestinal lining. With wide and attractive advertising campaigns, Merck succeeded in selling over 100 million prescriptions in the first year of the medication introduction.[72] In November 2000, The New England Journal of Internal Medicine (NEJM) published The VIGOR Clinical Trial.[73] A Randomized Controlled Trial assessing the risk of gastrointestinal bleeding in patients suffering from Rheumatoid Arthritis. The study randomized 8076 patients to receive either Naproxen or Rofecoxibe. The results of the clinical trial confirmed the suggested pathophysiological rationale. It showed that patients assigned to receive Vioxx had a lower incidence of gastrointestinal events compared to those who received Naproxen (2.1 in 100 patient-years in Vioxx arm compared to 4.5 in 100 patient-years, RR 0.5; 95% confidence interval 0.3 to 0.6; P value < 0.001). Unfortunately, the study also reported a higher incidence of Myocardial Infarction (MI) in patients receiving Roficoxibe (0.4% in the Vioxx arm, compared to 0.1% in the Naproxen arm).[74] The authors explained the latest observation by arguing that the findings might have been caused by the antiplatelet effect of Naproxen, citing a European study showing a similar protective effect using another NSAID—Flurbiprofen.[75] They also added that 38% of the patients who had a myocardial infarction (MI) in the study met the FDA criteria for using Aspirin for secondary prevention but were not actively taking the medication. In the following years, after publishing the VIGOR trial, many scholars voiced concerns about the cardiovascular safety profile of Rofecoxibe. In August 2001, Mukherjee and his colleagues published a review article in The Journal of American Medical Association (JAMA) discussing the results of four trials, including the VIGOR trial. After analyzing the study results with additional data, which was submitted later to the FDA by the study sponsor, the relative rate of sustaining cardiovascular events was higher in Vioxx group compared to Naproxen (RR 2.38; CI:1.39 to 4.00, P < 0.002).[76] The quarrel continue for four years until Rofecoxibe was voluntary withdrew from the market as a consequence of The APPROVe trial which showed higher incidence of thrombotic events in Vioxx group compared to Placebo (1.5 event per 100 patient-year in Vioxx compared to 0.78 event per 100 patient-year in Placebo, RR 1.92, 95% confidence interval, 1.19 to 3.11; $P = 0.008$) confirming the feared concerns.[77] It is roughly estimated that 88,000 patients in the united states suffered myocardial infarctions while taking Rofecoxibe.[78]

Since 1996, scientists at Merck knew that Vioxx reduced the levels of protective prostacyclins in the urine of healthy volunteers, and this could increase the tendency of having clotting events. Despite the early suspicions surrounding the safety of Vioxx, initial studies were not designed to primarily evaluate the cardiovascular risks (71). In fact, years later, when the picture became more evident, many questions were aired regarding how the pharmaceutical company, the authors, and NEJM handled the evidence and the scientific findings. First, the risk of cardiovascular events in VIGOR trial was not published in an absolute number of events but rather as a relative risk compared to Naproxen in a possible attempt to obscure the alarming findings. Second, three more patients suffered from Myocardial events in the last month of the trial, and those were not reported in the published manuscript. Third, there were some concerns about the presence of a conflict of interest, as two authors were employed by Merck. Moreover, the head of the study received a 2-year consulting contract from the pharmaceutical company and disclosed a family financial ownership reaching 70,000$.[79] Additionally, it took NEJM about a year from the time of Vioxx withdrawal (September 2004) to publish an expression of concern regarding the VIGOR study results. This delay was explained later by The Wall Street Journal as an attempt to preserve NEJM's reputation. Also, more than 900,000 copies of the VIGOR study were re-sold, generating significant revenue for the Journal. Dr. David Graham, an associate director in the FDA, commented on this story as "the worst preventable health disaster in the FDA's history."[80]

3.5.2   Definition and Analysis

Conflict of interest is defined as "a set of conditions in which an unbiased observer would determine that the judgment of a professional individual was at risk of being unduly influenced by his personal interest."[81] We commonly hear about this concept, especially in health care systems. In fact, people repeatedly refer to such circumstances with a sense of suspicion and doubt. However, it is not as easy as it seems to grasp the concept thoroughly. In the next section, I will try to break down the definition and offer a philosophical analysis of the notion. I will also briefly review the historical basis of the conflict of interest in clinical research. Finally, I will discuss some of the practiced and suggested safeguards in managing the potential hazards that might result from conflict of interest.

The legislation of the Bay-Dole act in 1980 was a major changing point in the history of scientific research in the United States.[82] Before this law the federal government held all the patents on scientific innovations attributed to federally funded research. As a result, many of these discoveries got warehoused and were not invested by the industry.[83] Not surprisingly, these scientific innovations had to pass through a "knowledge filter" of extreme government bureaucracy and regulations, which indirectly limited the options for investments, commercialization, and consequently, economic growth. The Bayh-Dole act allowed universities to hold patents on scientific discoveries and to form direct relationships with industry, which facilitated the movement of scientific discoveries from bench to market. Before the Bayh-Dole act, about fifty percent of scientific research was funded by the federal government, now more than 60% is funded by private industry.[84] Although these strong ties with private companies facilitated the exploitation of scientific fruits, it also raised my questions and concerns regarding the presence of a conflict of interests.[85]

Now, after this brief historical review, let us try to dissect the concept explicitly. The definition of a conflict of interest entails three elements: Primary Interest, Secondary Interest, and a Conflict. In health care and scientific research, the primary interest of the investigator is truth. He should strive to preserve the integrity of his work in order to generate accurate results. On the other side, in medical practice, the primary interest of a clinician is his patient's well-being. He should put aside all other interests and prioritize the patient's health and needs.[86] Theoretically, this is how things should work. However, life is much more complicated. The physician also has secondary interests in his private and professional life. For example, his family commitments, his financial stability, job reputation and professional growth. All of these interests might sometimes get in the way of professional conduct and contribute to generating conflicts. Usually, when people discuss secondary interests, they mainly refer to financial ones. There are two reasons for that: First, it is easier to quantify and understand funds. Second, it is generally agreed that financial gains should not be prioritized over professional judgments. In contrast, reputations, job positions, and other personal interests could sometimes be considered a competing primary interest.[87] So, the conflict arises when someone can prioritize one interest over another. When two interests are equal in importance and necessity, the notion is probably better described as a conflict of commitments rather than interests.[88]

Now, let us dive deeper into the concept. Consider the following analogy. Imagine that a hypothetical person named Jack went to a small electronic store in his neighborhood to buy a computer. One week later, after purchasing the new XYZ device, Jack was informed by his friend that the owner of that small store had profit contract with the XYZ company, and he usually tries to promote and sell their products for personal benefits. After learning about this fact, Jack might or might not feel annoyed or betrayed. He might pause for a second and reflect on his encounter with the seller, what he told him, and why did he end up choosing this computer brand. He might think he could have been deceived, but overall, it is the market, so things like that are expected. Now imagine a similar story happening between Jack and his physician. However, the recommended product is not a computer but rather a medication. How would jack feel if he knew that his physician had prescribed XYZ medication to him because he has financial ties with a pharmaceutical company?

This analogy takes us to the main point; at its essence, conflict of interest is a question of fidelity.[89] The relationship between a physician and a patient is grounded on trust, and any possible interference with this fidelity would wound and erode this relationship.[90] The patient or a research subject is a vulnerable human putting himself in the hands of a professional, and that is why conflict of interests are commonly mentioned in the fields which require sincere trust, like law, ministry and medicine.[91] As the Christian theologian Paul Ramsey once said, the fundamental ethical question lies in the meaning of faithfulness in our human relationships.[92]

The second important notion in the definition of conflict of interest is the word Risk. The dilemma is not in the evidence that something went wrong but rather in the tendency and the potential.[93] If a violation of trust or exploitation of power had proven to occur, then the situation should no longer be placed in the category of conflict of interest. It had become something else, like negligence or bribery. Any attempt to find hardcore evidence of a causative relationship totally misses the point. As Brody put it, imagine that a judge was assigned to a case of a lawsuit between two companies. And it was later discovered that she has a large-cap of stocks in one of these companies. How would people react to her judgment? Even if she ended up condemning the company in which she shared stocks, it would be hard to predict how her sentence would have differed if she did not have any financial interests. It could have been even more intense. There is no answer to this question.[94] The main problem lies in the possibility of a subconscious -unduly interference and not in providing evidence that a violation had occurred. In the field of clinical research, it is well known that the results of industry-sponsored clinical trials tend to favor the sponsoring companies compared to neutrally conducted trials.[95] Now, the next question to be asked: is there any biological evidence to back up the concept of the conflict of interest? What can we learn from moral psychology?

The field of moral psychology studies how our mind makes ethical decisions.[96] Wagar and Thagard developed the GAGE model to illustrate the way we integrate our emotional and cognitive functions during the decision-making process.[97] This model was inspired by the unfortunate case of Phineas Gage, a railway construction foreman who sustained a devastating brain injury by an iron rode during an accidental explosion.[98] The accident damaged his frontal lobe leaving him unable to reason and his personality changed to an indulgent-irresponsible one. According to the GAGE model, the decision-making process in the brain requires the integration of conscious and subconscious functions. The VMPFC (Ventero-Medial Pre-Frontal Cortex) connects the mainly cognitive pre-frontal cortex with the more emotional centers like the Amygdala, Nucleus Accumbens, and Hippocampus (memory center). Studies on gamblers showed that those who had VMPFC damage were unable to learn from their previous wrong decisions and continued to pick from bad decks and make wrong judgments.[99]

Psychologically, there are two affective afflictions involved in the conflict of interests: weakness of will and self-deception. Regarding the first, each one of us has a weakness or a flaw regarding a physical or mental stimulus, like food, drugs, power, sex, etc. While many times, someone might try to resist an impulse by rationing through his pre-frontal mathematical areas, his Nucleus Accubens and other parts of the limbic system would urge him to move against his rational will. Examples of such circumstances are quite common, like being unable to resist a good ice cream while on a diet. The second important affliction is self-deception. It is a subconscious process in which we try to convince ourselves that our decisions are moral to maintain our self-esteem regarding a subjective life's vision. An example would be a leader who tries to convince himself that he is a good person despite getting bribed.[100] The main point here is that our emotions play an essential role in the decision-making process. And a major part of this process is subconscious and uncontrolled.

Lastly, how can we control conflict of interest? If possible, someone should try to avoid any situation in which a conflict of interest might arise.[101] Many times, this is unfeasible, yet it is still advisable. Disclosure of any financial interest is another tool and it is recommended at present by most academic journals before publishing research studies.[102] The NIH also asks investigators to disclose significant financial interests before applying for contracts or grants. Management is another tool.[103] For example, moving a principal investigator with a significant financial interest into a consultant position or not allowing a physician with a potential conflict of interest to obtain informed consent. Finally, some conflict of interests needs to be totally prohibited, like those which lead to compromising the integrity of the research results, falsifying information, or those which give the sponsor full control over data analysis and without allowing academic staff to access or analyze the research results.[104]

3.6 Summary

The essential point to emphasize at the end of this chapter is the fundamental ethical tension between the conduct of clinical research and the individual patient's care. The former is mainly grounded on utilitarian theory, whereas the latter is primarily deontological. Many ethical issues might face the physician–investigator in trying to reconcile those two roles. Understanding the philosophical and ethical basis of each of these dilemmas is essential for good practice. Although sometimes the conduct of a clinical trial might compromise individual care for the futuristic benefit of others, the physician-scientist must strive to respect her patient's fidelity and vulnerability and maintain an ethical practice. By avoiding conflict of interest when possible, helping her patients make an informed decision, and ensuring that each patient receives the best available treatment option in the light of our knowledge and wisdom limitations as human beings.

Thought Box (III)
At the end of this chapter, it is necessary to elaborate a bit more on the concept of informed decision. I frequently get asked by many of my colleagues at work to start a research project to fine-tune the process of shared decision-making. For example, the question would be like the following: how can we ensure that a 50-year-old patient choosing LVAD over a heart transplant is making an informed decision? Can we build a model to help advance heart failure patients in the decision-making process? Don't you think our way of handling those sorts of discussions is quite paternalistic?

I have no doubts that these questions are all very valid ones. And I don't question the good character of my colleagues and many other physicians who strive to deliver the best possible care to their patients. However, I believe some notions need to be emphasized before discussing the feasibility of building a model of decision-making.

In the United States, there is a clear focus on the principle of patients' autonomy. We often hear statements such as "I own my body, I want this to be done, and not that treatment or device, etc." Yet as I have discussed before, autonomy should not be viewed as a power against the potential evil shadow of a good physician. Undoubtedly, the rise of the concept was, to some degree, a response to the exploitation of power differences between physicians and patients. However, there is no escape from the responsibility of a physician to guide a patient toward what is best for her in light of her values—knowing that there will always be decisions with uncertainties and there will always be a role for the good character of a physician in maintaining beneficence. No matter how educated our patients become, there will always be a knowledge gap, an experience difference that requires the physician's voice. When we study, discuss, and research, we try to shrink that space of uncertainty in which we perform our daily service. Yet, no matter how much we learn, there will always be a gap. No artificial intelligence model will be able to close it. And fully informed decision might not and will most likely not be attainable any time soon. Neither from a patient's standpoint nor the physician's standpoint. Our duty is to keep fidelity with our patients, work hard to expand our knowledge, keep a peaceful consciousness away from secondary gains, and admit our ignorance when we are ignorant. Building better and better models to help our patients is always necessary and required yet aiming to achieve a fully informed autonomous decision of our patients is just a delusion.

Notes

1. Halabi, Susan, and Stefan Michiels. "Introduction to Clinical Trials." *Textbook of Clinical Trials in Oncology: A Statistical Perspective* (2019). P 1.
2. Thompson, R. Paul, and Ross E.G. Upshur. *Philosophy of medicine: An introduction.* Routledge, 2017. P 5.
3. Nellhaus, Emma M., and Todd H. Davies. "Evolution of Clinical Trials throughout History." *Marshall Journal of Medicine* 3, no. 1 (2017): P 44.
4. Thompson, R. Paul, and Ross E.G. Upshur. *Philosophy of medicine: An introduction.* Routledge, 2017. Page 6.
5. Mitscherlich, Alexander, and Fred Mielke. *Doctors of infamy: the story of the Nazi medical crimes.* Pickle Partners Publishing, 2015. P 36–37.
6. Mitscherlich, Alexander, and Fred Mielke. *Doctors of infamy: the story of the Nazi medical crimes.* Pickle Partners Publishing, 2015. P 38–39.
7. Mitscherlich, Alexander, and Fred Mielke. *Doctors of infamy: the story of the Nazi medical crimes.* Pickle Partners Publishing, 2015. P 40–55.
8. Emanuel, Ezekiel J., Christine C. Grady, Robert A. Crouch, Reidar K. Lie, Franklin G. Miller, and David D. Wendler, eds. *The Oxford textbook of clinical research ethics.* Oxford University Press, 2008. P 23–24.
9. Kelly, David F., Gerard Magill, and Henk Ten Have. *Contemporary Catholic health care ethics.* Georgetown University Press, 2013. P 260–261.
10. Emanuel, Ezekiel J., Christine C. Grady, Robert A. Crouch, Reidar K. Lie, Franklin G. Miller, and David D. Wendler, eds. *The Oxford textbook of clinical research ethics.* Oxford University Press, 2008. P 88–89.
11. Barrett, Laura A. "Tuskegee Syphilis Study of 1932–1973 and the Rise of Bioethics as Shown Through Government Documents and Actions." *DttP: Documents to the People* 47, no. 4 (2019): 12.
12. Emanuel, Ezekiel J., Christine C. Grady, Robert A. Crouch, Reidar K. Lie, Franklin G. Miller, and David D. Wendler, eds. *The Oxford textbook of clinical research ethics.* Oxford University Press, 2008. P 90–93.
13. Kelly, David F., Gerard Magill, and Henk Ten Have. *Contemporary Catholic health care ethics.* Georgetown University Press, 2013. P 262–263.
14. Barrett, Laura A. "Tuskegee Syphilis Study of 1932–1973 and the Rise of Bioethics as Shown Through Government Documents and Actions." *DttP: Documents to the People* 47, no. 4 (2019): 13.
15. Beauchamp, Tom L., and James F. Childress. *Principles of biomedical ethics.* Oxford University Press, USA, 2001. P 333.
16. Kelly, David F., Gerard Magill, and Henk Ten Have. *Contemporary Catholic health care ethics.* Georgetown University Press, 2013. P 259.
17. Beauchamp, Tom L., and James F. Childress. *Principles of biomedical ethics.* Oxford University Press, USA, 2001. P 333.
18. Kelly, David F., Gerard Magill, and Henk Ten Have. *Contemporary Catholic health care ethics.* Georgetown University Press, 2013. P 259.
19. Arter, Joshua D. "The killing joke: why Batman doesn't kill the Joker." (2015). Page 38.
20. Beauchamp, Tom L., and James F. Childress. *Principles of biomedical ethics.* Oxford University Press, USA, 2001. P 362.

21. Pojman, Louis P., and James Fieser. *Cengage advantage ethics: Discovering right and wrong*. Nelson Education, 2017. P 123.
22. Pojman, Louis P., and James Fieser. *Cengage advantage ethics: Discovering right and wrong*. Nelson Education, 2017. P 127.
23. Kelly, David F., Gerard Magill, and Henk Ten Have. *Contemporary Catholic health care ethics*. Georgetown University Press, 2013. P 67.
24. Johnson, Robert, and Adam Cureton. "Kant's moral philosophy." (2004). Section 4 (Categorical and Hypothetical Imperatives).
25. Pojman, Louis P., and James Fieser. *Cengage advantage ethics: Discovering right and wrong*. Nelson Education, 2017. P 128.
26. Pojman, Louis P., and James Fieser. *Cengage advantage ethics: Discovering right and wrong*. Nelson Education, 2017. P 129–130.
27. Pojman, Louis P., and James Fieser. *Cengage advantage ethics: Discovering right and wrong*. Nelson Education, 2017. P 135.
28. Donaldson, Chase M. "Using Kantian ethics in medical ethics education." *Medical Science Educator* 27, no. 4 (2017): 841.
29. Beauchamp, Tom L., and James F. Childress. *Principles of biomedical ethics*. Oxford University Press, USA, 2001. P 363–364.
30. Pojman, Louis P., and James Fieser. *Cengage advantage ethics: Discovering right and wrong*. Nelson Education, 2017. P 103.
31. Beauchamp, Tom L., and James F. Childress. *Principles of biomedical ethics*. Oxford University Press, USA, 2001. P 355.
32. Pojman, Louis P., and James Fieser. *Cengage advantage ethics: Discovering right and wrong*. Nelson Education, 2017. P 104–105.
33. Beauchamp, Tom L., and James F. Childress. *Principles of biomedical ethics*. Oxford University Press, USA, 2001. P 357–358.
34. Kelly, David F., Gerard Magill, and Henk Ten Have. *Contemporary Catholic health care ethics*. Georgetown University Press, 2013. P 70–71.
35. Emanuel, Ezekiel J., ed. 2003. *Ethical and Regulatory Aspects of Clinical Research: Readings and Commentary*. Baltimore: Johns Hopkins University Press. P 117–120.
36. London, Alex John. "Equipoise in research: integrating ethics and science in human research." *Jama* 317, no. 5 (2017): 525–526.
37. Emanuel, Ezekiel J., Christine C. Grady, Robert A. Crouch, Reidar K. Lie, Franklin G. Miller, and David D. Wendler, eds. *The Oxford textbook of clinical research ethics*. Oxford University Press, 2008. P 253.
38. Bartlett, Robert H., Dietrich W. Roloff, Richard G. Cornell, Alice French Andrews, Peter W. Dillon, and Joseph B. Zwischenberger. "Extracorporeal circulation in neonatal respiratory failure: a prospective randomized study." *Pediatrics* 76, no. 4 (1985): 479–487.
39. O'Rourke, P. Pearl, Robert K. Crone, Joseph P. Vacanti, James H. Ware, Craig W. Lillehei, Richard B. Parad, and Michael F. Epstein. "Extracorporeal membrane oxygenation and conventional medical therapy in neonates with persistent pulmonary hypertension of the newborn: a prospective randomized study." *Pediatrics* 84, no. 6 (1989): 957–963.
40. Emanuel, Ezekiel J., ed. 2003. *Ethical and Regulatory Aspects of Clinical Research: Readings and Commentary*. Baltimore: Johns Hopkins University Press. P 122–123.
41. Emanuel, Ezekiel J., Christine C. Grady, Robert A. Crouch, Reidar K. Lie, Franklin G. Miller, and David D. Wendler, eds. *The Oxford textbook of clinical research ethics*. Oxford University Press, 2008. P 251.
42. Gelfand, Scott D. "A partial defense of clinical equipoise." *Research Ethics* 15, no. 2 (2019): Page 1.
43. Lo, Bernard. *Ethical issues in clinical research: A practical guide*. Lippincott Williams & Wilkins, 2012. P 243.
44. Emanuel, Ezekiel J., ed. 2003. *Ethical and Regulatory Aspects of Clinical Research: Readings and Commentary*. Baltimore: Johns Hopkins University Press. P 119.

45. Hey, Spencer Phillips, and Robert D. Truog. "The question of clinical equipoise and patients' best interests." *AMA journal of ethics* 17, no. 12 (2015): 1108–1109.
46. Gelfand, Scott D. "A partial defense of clinical equipoise." *Research Ethics* 15, no. 2 (2019): Page 4.
47. Emanuel, Ezekiel J., Christine C. Grady, Robert A. Crouch, Reidar K. Lie, Franklin G. Miller, and David D. Wendler, eds. *The Oxford textbook of clinical research ethics.* Oxford University Press, 2008. P 248–250.
48. Emanuel, Ezekiel J., ed. 2003. *Ethical and Regulatory Aspects of Clinical Research: Readings and Commentary.* Baltimore: Johns Hopkins University Press. P 120.
49. Gelfand, Scott D. "A partial defense of clinical equipoise." *Research Ethics* 15, no. 2 (2019): Page 8.
50. Emanuel, Ezekiel J., David Wendler, and Christine Grady. "What makes clinical research ethical?." *Jama* 283, no. 20 (2000): Page 2705–2706.
51. Jütte, Robert. "The early history of the placebo." *Complementary therapies in medicine* 21, no. 2 (2013): 94–97.
52. Krol, Fas Jacob, Michal Hagin, Eduard Vieta, Rephael Harazi, Amit Lotan, Rael D. Strous, Bernard Lerer, and Dina Popovic. "Placebo—To be or not to be? Are there really alternatives to placebo-controlled trials?." *European Neuropsychopharmacology* (2020). Page 2.
53. Lo, Bernard. *Ethical issues in clinical research: A practical guide.* Lippincott Williams & Wilkins, 2012. P 244–245.
54. Millum, Joseph, and Christine Grady. "The ethics of placebo-controlled trials: methodological justifications." *Contemporary clinical trials* 36, no. 2 (2013): 510–514.
55. Emanuel, Ezekiel J., ed. 2003. *Ethical and Regulatory Aspects of Clinical Research: Readings and Commentary.* Baltimore: Johns Hopkins University Press. P 142.
56. Temple, Robert, and Susan S. Ellenberg. "Placebo-controlled trials and active-control trials in the evaluation of new treatments. Part 1: ethical and scientific issues." *Annals of internal medicine* 133, no. 6 (2000): 455–463.
57. Hey, Spencer Phillips, and Charles Weijer. "Assay sensitivity and the epistemic contexts of clinical trials." *Perspectives in biology and medicine* 56, no. 1 (2013): 1–17. (Page 7).
58. Emanuel, Ezekiel J., ed. 2003. *Ethical and Regulatory Aspects of Clinical Research: Readings and Commentary.* Baltimore: Johns Hopkins University Press. P 136.
59. Emanuel, Ezekiel J., ed. 2003. *Ethical and Regulatory Aspects of Clinical Research: Readings and Commentary.* Baltimore: Johns Hopkins University Press. P 129 and P 141.
60. Pope, Thaddeus Mason. "Certified patient decision aids: Solving persistent problems with informed consent law." *The Journal of Law, Medicine & Ethics* 45, no. 1 (2017): 12–40. Page 14.
61. Emanuel, Ezekiel J., ed. 2003. *Ethical and Regulatory Aspects of Clinical Research: Readings and Commentary.* Baltimore: Johns Hopkins University Press. P 197.
62. Emanuel, Ezekiel J., Christine C. Grady, Robert A. Crouch, Reidar K. Lie, Franklin G. Miller, and David D. Wendler, eds. *The Oxford textbook of clinical research ethics.* Oxford University Press, 2008. P 606–607.
63. Sulmasy, Daniel P. *The rebirth of the clinic: An introduction to spirituality in health care.* Georgetown University Press, 2006. P 32–33.
64. Pellegrino, Edmund D. "Toward a reconstruction of medical morality." *The American Journal of Bioethics* 6, no. 2 (2006): 65–71.
65. Emanuel, Ezekiel J., ed. 2003. *Ethical and Regulatory Aspects of Clinical Research: Readings and Commentary.* Baltimore: Johns Hopkins University Press. P 203–204.
66. Emanuel, Ezekiel J., Christine C. Grady, Robert A. Crouch, Reidar K. Lie, Franklin G. Miller, and David D. Wendler, eds. *The Oxford textbook of clinical research ethics.* Oxford University Press, 2008. P 608–609.
67. Biros, Michelle. "Capacity, vulnerability, and informed consent for research." *The Journal of Law, Medicine & Ethics* 46, no. 1 (2018): 72–78. Page 75.
68. Emanuel, Ezekiel J., ed. 2003. *Ethical and Regulatory Aspects of Clinical Research: Readings and Commentary.* Baltimore: Johns Hopkins University Press. P 206–207.

69. Biros, Michelle. "Capacity, vulnerability, and informed consent for research." *The Journal of Law, Medicine & Ethics* 46, no. 1 (2018): 72–78. Page 72.

70. Emanuel, Ezekiel J., ed. 2003. *Ethical and Regulatory Aspects of Clinical Research: Readings and Commentary*. Baltimore: Johns Hopkins University Press. P 200.

71. Krumholz, Harlan M., Joseph S. Ross, Amos H. Presler, and David S. Egilman. "What have we learnt from Vioxx?." *Bmj* 334, no. 7585 (2007): 120.

72. McIntyre, William F., and Gerald Evans. "The Vioxx® legacy: Enduring lessons from the not so distant past." *Cardiology journal* 21, no. 2 (2014): 203.

73. Bombardier, Claire, Loren Laine, Alise Reicin, Deborah Shapiro, Ruben Burgos-Vargas, Barry Davis, Richard Day et al. "Comparison of upper gastrointestinal toxicity of rofecoxib and naproxen in patients with rheumatoid arthritis." *New England Journal of Medicine* 343, no. 21 (2000): 1520–1528.

74. Lo, Bernard. *Ethical issues in clinical research: A practical guide*. Lippincott Williams & Wilkins, 2012. P 134.

75. Brochier, M. L. "Evaluation of flurbiprofen for prevention of reinfarction and reocclusion after successful thrombolysis or angioplasty in acute myocardial infarction." *European heart journal* 14, no. 7 (1993): 951–957.

76. Mukherjee, Debabrata, Steven E. Nissen, and Eric J. Topol. "Risk of cardiovascular events associated with selective COX-2 inhibitors." *Jama* 286, no. 8 (2001): 954–959.

77. Bresalier, Robert S., Robert S. Sandler, Hui Quan, James A. Bolognese, Bettina Oxenius, Kevin Horgan, Christopher Lines et al. "Cardiovascular events associated with rofecoxib in a colorectal adenoma chemoprevention trial." *New England Journal of Medicine* 352, no. 11 (2005): 1092–1102.

78. McIntyre, William F., and Gerald Evans. "The Vioxx® legacy: Enduring lessons from the not so distant past." *Cardiology journal* 21, no. 2 (2014): 204.

79. Krumholz, Harlan M., Joseph S. Ross, Amos H. Presler, and David S. Egilman. "What have we learnt from Vioxx?." *Bmj* 334, no. 7585 (2007): 120.

80. Wilson, M. "The New England Journal of Medicine: commercial conflict of interest and revisiting the Vioxx scandal." *Indian journal of medical ethics* 1, no. 3 (2016): 167–171.

81. Beauchamp, Tom L., and James F. Childress. *Principles of biomedical ethics*. Oxford University Press, USA, 2001. P 328.

82. Lo, Bernard. *Ethical issues in clinical research: A practical guide*. Lippincott Williams & Wilkins, 2012. P 132.

83. Aldridge, T. Taylor, and David Audretsch. "The Bayh-Dole act and scientist entrepreneurship." In *Universities and the Entrepreneurial Ecosystem*. Edward Elgar Publishing, 2017.

84. Emanuel, Ezekiel J., Christine C. Grady, Robert A. Crouch, Reidar K. Lie, Franklin G. Miller, and David D. Wendler, eds. *The Oxford textbook of clinical research ethics*. Oxford University Press, 2008. P 758.

85. Emanuel, Ezekiel J., ed. 2003. *Ethical and Regulatory Aspects of Clinical Research: Readings and Commentary*. Baltimore: Johns Hopkins University Press. P 369.

86. Steinbrook, Robert. "Controlling conflict of interest—proposals from the Institute of Medicine." *New England Journal of Medicine* 360, no. 21 (2009): 2160–2163.

87. Emanuel, Ezekiel J., Christine C. Grady, Robert A. Crouch, Reidar K. Lie, Franklin G. Miller, and David D. Wendler, eds. *The Oxford textbook of clinical research ethics*. Oxford University Press, 2008. P 760.

88. Emanuel, Ezekiel J., Christine C. Grady, Robert A. Crouch, Reidar K. Lie, Franklin G. Miller, and David D. Wendler, eds. *The Oxford textbook of clinical research ethics*. Oxford University Press, 2008. P 761.

89. Fineberg, Harvey V. "Conflict of interest: why does it matter?." *Jama* 317, no. 17 (2017): 1717–1718.

90. Beauchamp, Tom L., and James F. Childress. *Principles of biomedical ethics*. Oxford University Press, USA, 2001. P 324–325.

91. Lemmens, Trudo, and Peter A. Singer. "Bioethics for clinicians: 17. Conflict of interest in research, education and patient care." *Cmaj* 159, no. 8 (1998): 961 (The concept of patient's vulnerability).
92. Ramsey, Paul, Margaret A. Farley, and Albert R. Jonsen. *The patient as person: explorations in medical ethics.* Yale University Press, 2002. P XLV.
93. Lo, Bernard. *Ethical issues in clinical research: A practical guide.* Lippincott Williams & Wilkins, 2012. P 133.
94. Brody, Howard. "Clarifying conflict of interest." *The American journal of bioethics* 11, no. 1 (2011): 24.
95. Brody, Howard. "Clarifying conflict of interest." *The American journal of bioethics* 11, no. 1 (2011): 25. Refer to the cited study by Lexchin et al. 2003.
96. Thagard, Paul. "The moral psychology of conflicts of interest: Insights from affective neuroscience." *Journal of Applied Philosophy* 24, no. 4 (2007): 369.
97. Thagard, Paul. "The moral psychology of conflicts of interest: Insights from affective neuroscience." *Journal of Applied Philosophy* 24, no. 4 (2007): 370–371.
98. O'Driscoll, Kieran, and John Paul Leach. ""No longer Gage": an iron bar through the head: Early observations of personality change after injury to the prefrontal cortex." (1998): 1673–1674.
99. Thagard, Paul. "The moral psychology of conflicts of interest: Insights from affective neuroscience." *Journal of Applied Philosophy* 24, no. 4 (2007): 372.
100. Thagard, Paul. "The moral psychology of conflicts of interest: Insights from affective neuroscience." *Journal of Applied Philosophy* 24, no. 4 (2007): 374–375.
101. Brody, Howard. "Clarifying conflict of interest." *The American journal of bioethics* 11, no. 1 (2011): 23–28.
102. Emanuel, Ezekiel J., Christine C. Grady, Robert A. Crouch, Reidar K. Lie, Franklin G. Miller, and David D. Wendler, eds. *The Oxford textbook of clinical research ethics.* Oxford University Press, 2008. P 764.
103. Lo, Bernard. *Ethical issues in clinical research: A practical guide.* Lippincott Williams & Wilkins, 2012. P 137.
104. Lo, Bernard. *Ethical issues in clinical research: A practical guide.* Lippincott Williams & Wilkins, 2012. P 138–139.

Bibliography

1. Aldridge, T. Taylor, and David Audretsch. 2017. The Bayh-Dole act and scientist entrepreneurship. In *Universities and the entrepreneurial ecosystem.* Cheltenham: Edward Elgar Publishing.
2. Arter, Joshua D. 2015. *The killing joke: Why Batman doesn't kill the Joker.*
3. Barrett, Laura A. 2019. Tuskegee syphilis study of 1932–1973 and the rise of bioethics as shown through government documents and actions. *DttP: Documents to the People* 47 (4): 11–16.
4. Bartlett, Robert H., Dietrich W. Roloff, Richard G. Cornell, Alice French Andrews, Peter W. Dillon, and Joseph B. Zwischenberger. 1985. Extracorporeal circulation in neonatal respiratory failure: A prospective randomized study. *Pediatrics* 76 (4): 479–487.
5. Beauchamp, Tom L., and James F. Childress. 2001. *Principles of biomedical ethics.* USA: Oxford University Press.
6. Biros, Michelle. 2018. Capacity, vulnerability, and informed consent for research. *The Journal of Law, Medicine & Ethics* 46 (1): 72–78, 75.
7. Bombardier, Claire, Loren Laine, Alise Reicin, Deborah Shapiro, Ruben Burgos-Vargas, Barry Davis, Richard Day, et al. Comparison of upper gastrointestinal toxicity of rofecoxib and naproxen in patients with rheumatoid arthritis. *New England Journal of Medicine* 343 (21): 1520–1528.

8. Bresalier, Robert S., Robert S. Sandler, Hui Quan, James A. Bolognese, Bettina Oxenius, Kevin Horgan, Christopher Lines, et al. 2005. Cardiovascular events associated with rofecoxib in a colorectal adenoma chemoprevention trial. *New England Journal of Medicine* 352 (11): 1092–1102.
9. Brochier, M.L. 1993. Evaluation of flurbiprofen for prevention of reinfarction and reocclusion after successful thrombolysis or angioplasty in acute myocardial infarction. *European Heart Journal* 14 (7): 951–957.
10. Brody, Howard. 2011. Clarifying conflict of interest. *The American Journal of Bioethics* 11 (1).
11. Donaldson, Chase M. 2017. Using Kantian ethics in medical ethics education. *Medical Science Educator* 27 (4): 841–845.
12. Emanuel, Ezekiel J., ed. 2003. *Ethical and regulatory aspects of clinical research: Readings and commentary*, 117–120. Baltimore: Johns Hopkins University Press.
13. Emanuel, Ezekiel J., ed. 2003. *Ethical and regulatory aspects of clinical research: Readings and commentary*. Baltimore: Johns Hopkins University Press.
14. Emanuel, Ezekiel J., Christine C. Grady, Robert A. Crouch, Reidar K. Lie, Franklin G. Miller, and David D. Wendler (eds.). 2008. *The Oxford textbook of clinical research ethics*. Oxford: Oxford University Press.
15. Emple, Robert, and Susan S. Ellenberg. 2000. Placebo-controlled trials and active-control trials in the evaluation of new treatments. Part 1: Ethical and scientific issues. *Annals of Internal Medicine* 133 (6): 455–463.
16. Fineberg, Harvey V. 2017. Conflict of interest: Why does it matter? *JAMA* 317 (17): 1717–1718.
17. Gelfand, Scott D. 2019. A partial defense of clinical equipoise. *Research Ethics* 15 (2): 1.
18. Halabi, Susan, and Stefan Michiels. 2019. Introduction to clinical trials. In *Textbook of clinical trials in oncology: A statistical perspective*.
19. Hey, Spencer Phillips, and Charles Weijer. 2013. Assay sensitivity and the epistemic contexts of clinical trials. *Perspectives in Biology and Medicine* 56 (1): 1–17, 7.
20. Hey, Spencer Phillips, and Robert D. Truog. 2015. The question of clinical equipoise and patients' best interests. *AMA Journal of Ethics* 17 (12): 1108–1109.
21. Johnson, Robert, and Adam Cureton. 2004. *Kant's moral philosophy*, sect. 4 (Categorical and hypothetical imperatives).
22. Jütte, Robert. 2013. The early history of the placebo. *Complementary Therapies in Medicine* 21 (2): 94–97.
23. Kelly, David F., Gerard Magill, and Henk Ten Have. 2013. *Contemporary Catholic health care ethics*. Washington, D.C.: Georgetown University Press.
24. Krol, Fas Jacob, Michal Hagin, Eduard Vieta, Rephael Harazi, Amit Lotan, Rael D. Strous, Bernard Lerer, and Dina Popovic. 2020. Placebo—To be or not to be? Are there really alternatives to placebo-controlled trials? *European Neuropsychopharmacology* 2.
25. Krumholz, Harlan M., Joseph S. Ross, Amos H. Presler, and David S. Egilman. 2007. What have we learnt from Vioxx? *BMJ* 334 (7585): 120.
26. Lemmens, Trudo, and Peter A. Singer. 1998. Bioethics for clinicians: 17. Conflict of interest in research, education and patient care. *CMAJ* 159 (8): 960–965.
27. Lo, Bernard. 2012. *Ethical issues in clinical research: A practical guide*. Philadelphia: Lippincott Williams & Wilkins.
28. London, Alex John. 2017. Equipoise in research: Integrating ethics and science in human research. *JAMA* 317 (5): 525–526.
29. McIntyre, William F., and Gerald Evans. 2014. The Vioxx® legacy: Enduring lessons from the not so distant past. *Cardiology Journal* 21 (2): 203–205.
30. Millum, Joseph, and Christine Grady. 2013. The ethics of placebo-controlled trials: Methodological justifications. *Contemporary Clinical Trials* 36 (2): 510–514.
31. Mitscherlich, Alexander, and Fred Mielke. 2015. *Doctors of infamy: The story of the Nazi medical crimes*. Auckland: Pickle Partners Publishing.
32. Mukherjee, Debabrata, Steven E. Nissen, and Eric J. Topol. 2001. Risk of cardiovascular events associated with selective COX-2 inhibitors. *JAMA* 286 (8): 954–959.

33. Nellhaus, Emma M., and Todd H. Davies. 2017. Evolution of clinical trials throughout history. *Marshall Journal of Medicine* 3 (1).
34. O'Driscoll, Kieran, and John Paul Leach. 1998. "No longer Gage": An iron bar through the head. Early observations of personality change after injury to the prefrontal cortex. *BMJ* 317 (7174): 1673–1674.
35. O'Rourke, P. Pearl, Robert K. Crone, Joseph P. Vacanti, James H. Ware, Craig W. Lillehei, Richard B. Parad, and Michael F. Epstein. Extracorporeal membrane oxygenation and conventional medical therapy in neonates with persistent pulmonary hypertension of the newborn: A prospective randomized study. *Pediatrics* 84 (6): 957–963.
36. Pellegrino, Edmund D. 2006. Toward a reconstruction of medical morality. *The American Journal of Bioethics* 6 (2): 65–71.
37. Pojman, Louis P., and James Fieser. 2017. Cengage advantage ethics: Discovering right and wrong. Toronto: Nelson Education.
38. Pope, Thaddeus Mason. 2017. Certified patient decision aids: Solving persistent problems with informed consent law. *The Journal of Law, Medicine & Ethics* 45 (1): 12–40, 14.
39. Ramsey, Paul, Margaret A. Farley, and Albert R. Jonsen. 2002. *The patient as person: Explorations in medical ethics.* London: Yale University Press.
40. Steinbrook, Robert. 2009. Controlling conflict of interest—Proposals from the Institute of Medicine. *New England Journal of Medicine* 360 (21): 2160–2163.
41. Sulmasy, Daniel P. 2006. *The rebirth of the clinic: An introduction to spirituality in health care,* 32–33. Washington, D.C.: Georgetown University Press.
42. Thagard, Paul. 2007. The moral psychology of conflicts of interest: Insights from affective neuroscience. *Journal of Applied Philosophy* 24 (4): 367–380.
43. Wilson, M. 2016. The New England Journal of Medicine: Commercial conflict of interest and revisiting the Vioxx scandal. *Indian Journal of Medical Ethics* 1 (3): 167–171.

Chapter 4
The Debate on Biotechnology and Genetic Therapy

Questions

- What do we mean by human dignity?
- What are the differences between therapy and enhancement in gene therapy?
- Do you think that the potential benefits will outweigh the associated risks in gene therapy?

4.1 Introduction

While studying in medical school, understanding the pathophysiology of various diseases was very interesting and exciting. For me, no class beat the physiology class. I still remember, and very well indeed, the colorful diagrams and the cause-and-effect algorithms of different disease processes. Bottom line, studying those books was, for me, simply a delight. A few years later, when I moved to practice clinical medicine, hospital rounding was different. Reality was much more complicated than the colorful diagrams in my physiology books. And the manifestations of various diseases were not as straightforward and predictable as I thought. If we want to be realistic, we should admit that modern medicine has changed the clinical course of many illnesses by understanding the molecular levels of different pathologies. Yet, we have learned over the last few decades that our bodies are much more complicated than we thought. The human genetic make-up is exceptionally delicate, and our genes probably have a role in every disease pathogenicity. With the recent advancements in gene therapy, new hopes have emerged. Hopes that maybe we would be able to prevent autoimmune diseases and correct bad genetic predispositions proactively. Imagine a society without atherosclerosis, obesity, cystic fibrosis, cancers, and Huntington's disease. What if we could also enhance our immunity to prevent deadly viral pandemics? What if we can change our genome to become more intelligent, stronger, more peaceful, and less violent beings? The power that could be achieved has no limits. However, on the other hand, what if such hypothetical power

S. Toro, *Introduction to Clinical Ethics: Perspectives from a Physician Bioethicist*,
https://doi.org/10.1007/978-3-031-30804-8_4

was used in a destructive way, such as bioterrorism? Should we at all be concerned with the progress of this technology?

4.2 General Background

The recent advances in scientific knowledge gave the human being an unprecedented power to change his daily life, and his surrounding world. Since the rise of the scientific revolution in the sixteenth and seventeenth centuries, many discoveries, breakthroughs, and innovations radically reshaped our daily lives. Indeed, the pace of these successes rapidly accelerated during the twentieth century, and it affected, directly and indirectly, various fields and domains. Medicine was one of them. It witnessed a plentitude of great innovations in the areas of diagnostics, biotechnology, pharmaceuticals, and many more. Perhaps, one of the greatest milestones in that journey of achievements was the discovery of the double helix DNA structure by Watson and Crick in 1953.[1] This success opened up the door to further understanding of our genetic makeup, phenotypic features, and different hereditary diseases with many hopes and wishes for feature treatments and therapeutic applications.

4.2.1 The Cell and Genome Editing

Inside each of our cells lies a 2-m-long DNA. This DNA is Folded at a ratio of 1000 to 10,000 times to fit in the spherical nucleus of less 10 μm in diameter.[2] The building block of the DNA is the nucleotide, which is composed of Deoxyribose, a Phosphate group, and one of four different Nitrogenous bases. Simply put, these Nitrogenous bases (Adenine, Guanine, Cytosine, and Thiamine) are the four letters of the entire genetic language.[3] Each sequence of three nucleotides forms the basic coding unit, the Codon. Codons are later translated to Amino acids, which combine together by peptide bonds to form various proteins. Proteins play the predominant role in cellular functions and body hemostasis. Four different protein structures are present in our biological system, primary, secondary, tertiary, and quaternary. The complexity of the protein molecules increased moving forward, starting from a primary protein, a single polypeptide molecule to a more complex α-helix and β-pleated sheet (secondary protein), reaching to three-dimensional structures supported by covalent and non-covalent interactions (tertiary and quaternary).[4] Many hereditary diseases are caused by abnormal DNA sequences (mutated genes), which subsequently leads to defective-malfunctioning proteins (Sickle Cell Disease, Thalassemia, Alpha-1 Antitrypsin deficiency).

Since discovering the DNA structure in the middle of the twentieth century, scientists worked hard on expanding our genetic knowledge and developed tools to intervene on various genetic defects. My goal here is not to provide an extensive historical review of the progress of genetic therapies; however, it is necessary to emphasize a few important milestones in this historical context. 1962, Waclaw Szybalski made a pioneering discovery. He showed that a genetic defect could be corrected by transferring a foreign genetic material into defective bone marrow cells. He isolated the DNA of HHRPT (Hypoxanthine–guanine phosphoribosyl transferase) Positive cells and transferred the genetic information into HHRPT Negative cells enabling the later to survive in the presence of Aminopterin (a compound that inhibits dihydrofolate reductase and Purine synthesis).[5] In December 1988, The Recombinant DNA Advisory Committee (RAC) approved for the first time a clinical protocol to introduce foreign genes into a human body. Rosenberg isolated tumor-infiltrating Lymphocytes from metastatic melanoma. He modified the lymphocytes ex vivo to express tumor necrosis factor and reintroduced them into two patients suffering malignant melanoma. The results of the study showed no growth of tumor cells at the injection sites. In 2003, China was the first country in the world to approve the use of genetic therapy pharmaceutical names (Gendicine) to treat head and neck squamous cell carcinoma.[6] In 2012, the EMA approved the use of (Glybera- Alipogene tiparvovec), a viral vector for treating sever Lipoprotein Lipase deficiency.[7] In the USA, the FDA had already approved some Gene therapies, for example, (tisagenlecleucel) for Large B cell lymphoma, (voretigene neparvovec-rzyl) for biallelic RPE65 mutation-associated retinal dystrophy, (talimogene laherparepvec) for malignant melanoma. Currently, the US Government Database lists more than 3000 ongoing clinical studies for gene therapies in 204 different countries.[8]

4.2.2 CRSPR and Future Dreams

Various technologies for manipulating and introducing changes to the DNA structure have been developed since the discovery of the double helix structure of the DNA. These technologies played a significant role in advancing the field of molecular biology and genetic engineering. Initially, researchers depended on Meganucleases, like I-Scel, to introduce the required changes to the DNA sequence. Those proteins recognized long stretches of DNA (14–40 bp), which made them imprecise tools in making DNA changes.[9] Subsequently, more accurate proteins were discovered and applied in biotechnology. Zinc fingers, for example, recognize 3-bp DNA and TALE (transcription activator-like effector) proteins recognize single bp (base pair). (9) Despite being more effective in the genetic engineering field, the main problem that researchers faced was the complexity of designing different proteins with different DNA-recognition domains.[10]

In 2012, researchers introduced the CRISPR Cas9 system as a revolutionary tool in genomic biotechnology. CRISPR Cas9 uses an RNA sequence to recognize specific sites on the host DNA and introduce a double-strand break to make the required change. CRISPR CAS9 was initially discovered as a bacterial defense mechanism against phage infections and plasmid transfers.[11] This biological complex consists of two parts: sgRNA and Cas9 endonuclease. The sgRNA is formed by two parts: the crRNA and trans-activating RNA. The 3' of sgRNA is the constant part which forms a scaffold and binds to the Cas9 endonuclease, and 5',which is a variable part complementary to the targeted DNA sequence.[12] When a foreign DNA (from Phage or Plasmid) enters the bacteria, a piece of this foreign DNA gets integrated into the CRISPR array. Following that, a complex process of transcription, and post transcription modification leads to the formation of a mature crRNA-Cas complex. These complexes function as a memory unit to defend the bacteria against subsequent future bacteriophage infections.[13]

The CISPR Cas9 system opened doors to many possible future applications as a new, easier to engineer and more efficient biotechnology. Not only in gene editing but also as a tool in modifying genetic expression and epigenetic markers.[14] Despite all these promises, many challenges and concerns remain on the table for discussion. Off-target cutting and challenges with different delivery systems are major ones.[15] Many scientists and bioethicists have also raised ethical concerns regarding the safety of this technology, especially if applied to germline cells with possible unanticipated risks, and many advocated that excitement should be balanced by caution, and enthusiasm should be guided by humility in order to avoid future risks.[16]

4.3   Grounding the Debate

At present, many ethical questions regarding the applications of biotechnology and genetic engineering come to our minds. TV shows, movies, and science fiction novels are all trying to paint a speculative picture of a promising yet somewhat concerning future. Before reviewing a few of the highly debated topics in genetic engineering, I believe it is necessary to comment on two essential notions: the concepts of human *dignity* and the theological vision regarding the *Sanctity* of the human body.

4.3.1   Human Dignity

In recent years, many scholars questioned the value of dignity as a reference point in ethical debates, given the vagueness of the concept and the lack of a generalized consensus about its definition.[17] Ruth Macklin's, for example, criticized the ambiguity of the notion and described it as a mere slogan that can be grounded completely on human's autonomy.[18] Steven Pinker, in an essay named *The Stupidity of Dignity*, criticized the religious roots of the concept and gave examples from our daily lives in which we voluntarily agree to many actions that violate our dignity, like having sexual intercourse, agreeing to a rectal/pelvic exam to simpler things like getting out of a small car.[19] However, despite lacking a standard definition, the notion still has a shared resonance in the collective consciousness of people. Indeed, many debated abstract topics are still valuable, despite not having a shared definition, such as Love and Freedom. Also, the claim of having roots in the religious landscape does not undermine the value of human Dignity. On the contrary, it can be said that it adds to it and enriches it. For example, Justice and Autonomy are both frequently discussed topics in theological literature, and that fact by itself does not decrease their importance and value in our society. According to Jurgen Haberman, Egalitarian Universalism and the idea of freedom and social solidarity are both essentially grounded on the Judo-Christian Theology of justice and love.[20] We still highly appreciate and value those notions despite their solid religious links.

So, what does human dignity mean? Sulmasy offered a deliberate breakdown and examination of the concept. His classification helps to clarify some of the ambiguity in its use in literature. He refers to three different uses of the word dignity. The first is *intrinsic*, which is the human being's value simply because he is a human being.[21] As I will be discussing later, this understanding of the notion is deeply rooted in Judo-Christian theology. However, it is also grounded on the secular grounds of humans' rationality, autonomy, and the special capacities that humans possess compared to other creatures. This intrinsic dignity is not a result of any human attribute, and it cannot be erased.

The second use of the word is *attributed* dignity. It is the value that someone gains by education, work, money, social worth, or position. Omanthuna refers to the etymology of the word (Dignitas), which still captures the meaning.[22] We usually discuss this use of dignity when we talk about suffering. Patients who have significant comorbidity, disabling strokes, on artificial nutrition, or those suffering from various conditions that limit their capacity to act autonomously might feel a lack of worth. We say their dignity is at threat.[23] This use of the word refers to a spectrum in which dignity can be increased or decreased. Compared to the first use of dignity, which has absolute meaning. Someone can accept or deny its presence, but it does not come in various ranges. There are no degrees to intrinsic dignity. Lastly, Sulmasy refers to the concept of *inflorescent* dignity, in which a person flourishes in acting with what is consistent with being a human. Dignity, in this sense, is not totally attributed to the act or the results of the act itself but also to the shared consensus about what constitutes human excellence. Dignity here entails a state of virtue. (21)

4.3.2 Theological Background

A religious reference is always inevitable when people debate the various applications of genetic engineering. Whether someone agrees or rejects the role of religion in articulating arguments, a reference to a transcendent relationship is always present in the context of the discussions. Our understanding of the notion of the intrinsic value of a human being is significantly grounded on a theological base. In Catholic moral theology, our dignity as humans is derived from two facts: our creation in the image of God and our redemption of sins in Jesus Christ.[24] In Hebrew, the words "Selem" and "Demuth" means "Image" and "Likeness," giving human beings a Devine status by grace and a special position in God's creation.[25] In a sense, this dignity that they possess is an Alien dignity, not grounded on what people think about each other, but directly granted to people by God himself. This dignity transcends

a person and transcends his ability to reject it.[26] It granted Humans a Daimonion
over nature and animals (Adam named animals). Unlike animals, humans have the
capacity to transcend who they are, reflect, judge, and try to change themselves and
their societies (25). They derived this authority from the grace of God. Nevertheless,
this special status in creation did not give humans absolute authority over the rest
of the creatures. Along with this power came a major responsibility to act as good
stewards of God's grace, so in a sense, humans are both creatures and co-creators,
invited to accept the grace of god and distended to a Devine life.[27] In Greek Orthodox
literature, we find similar notions with the "sacredness" of human beings forming the
basic element of orthodox anthropology and *Theosis* being the telos of the Christian
life, to become "after god's likeness."[28]

So how are we supposed to answer the question of whether our human nature is
open to self-manipulation? Are we invited to accept our limitations and suffering, or
should we try to overcome them? Unfortunately, there is no clear answer to this question. As we will see in the context of our discussion, the relationship between God's
grace and human nature is extremely complex, and the theological principles that
ground this relationship, namely, the Divine Sovereignty and Redemptive Suffering,
cuts in both ways. They do not offer an easy way out or a clear yes/no answer to
whether we should accept the manipulations of our own human nature.[29] Before
1960, during the ethical debates around technological advancements, the theological
themes of Divine sovereignty and redemptive suffering were secondary to strong
physicalism and positivism in catholic moral theology. If a procedure was deemed
acceptable by the church, then the results were congruent with the Divine sovereignty
of a human being; if not, then he was invited to accept suffering as redemptive
(examples IVF, direct sterilization, which according to the catholic church are deontologically wrong).[30] Similar paradoxical notions in daily life can be found even
in secular terminologies (nature vs. nurture) or (optimism vs. pessimism about life,
which theologically mirrors the image of fall and creation).[31]

Karl Rahner (Catholic theologian) wrote two essays in the early 1960s in which
he examined whether human nature is opened or closed to self-manipulation. In the
first essay, Rahner was more positive about the idea. Although he refused to accept
all the possible technological methods, he thought that Christians should be ready to
accept future changes. Self-manipulation for him was neither a guarantee to hell nor
heaven. Moreover, in his opinion, no matter how far technology went, humans would
not succeed in changing their essence of determining good and evil.[32] In the second
essay, Rahner took a more conservative and skeptical position. He wrote that human
nature is predetermined, and people should accept it and not try to change it. Rahner
emphasized that man has not called himself into existence, and he cannot change
anything without using something already created. He saw that human nature was
so open and free to the point that any further attempt to increase its capacity might
lead to paradoxical effects.[33]

Whether should a Christian accept or reject genetic manipulation is a difficult
question. However, though not being able to offer a clear-cut answer, theological
principles can give us a spirit through which we can carry on the discussion. They
can remind us that there are no easy shortcuts, that human nature is mysterious and

sacred, and we should be extremely cautious about how to use our freedom and knowledge.[34]

Lastly, it is worth commenting on the possible role of the principle of proportionality in addressing the debate around human genomic manipulation. Peter Knauer wrote suggesting that proportionate reasons should balance the acts of evil. In a sense, the principle is consequentialist, but there is also a focus on the drives and the intentions of the person and not only on the results of the act. Since having bad intentions and being less virtuous would likely increase the probability of someone acting in an immoral way.[35] In Greek Orthodox theology, we find similar reference to the importance of the virtue of *Diacrisis* (discernment) in guiding moral decisions. Baloyannis cautioned that the motives behind human enhancements should not be coming from dissatisfaction or vanity and emphasized the role of conscious in guiding the moral action.[36]

4.4 Eugenics and Future Challenges

Eugenics still has a negative connotation at present. The word itself continues to bring a sense of u angst in public discourse. Whether the whole idea of eugenics is intrinsically bad is a thorny question to answer. Indeed, it depends on how we define the concept, what is our shared vision of a "better" future, and which means are appropriate to achieve that targeted future?

The rise of biotechnology and genetic engineering in the last few decades brought back the utopian dreams of Francis Galton; to improve our "human stock," to remove those deemed unfit from the world, and to control what nature blindly and recklessly offered us.[37] Advocates for the new eugenics claim that they have diagnosed the main problem of old eugenics, that their liberal, non-coercive, non-government driven approach will set new eugenics free from the old charges and will put us on a new path of perusing what is good for our children and future generations. There will be no more forced sterilizations of mental patients and prisoners, no contests of fitter families, and NO WAY Nazi-style racial hygiene and mass murder.[38] On the other side, Skeptics about this new approach argue that the new eugenics have the same spirit of old eugenics. However, instead of being forced by the government, it will be guided by the free-market, which brings its own concerns and problems.[39] The question that remains is the following: Are we allowed to shape our children the way we want? Aren't we already practicing eugenics by engaging our children in various sports, or by offering them a healthy diet? What about choosing the best schools to offer excellent education? Aren't all those decisions eugenics ones?

4.4.1   The New Eugenics

Charles Darwin wrote that welfare programs, despite being a natural expression of caring and empathy, are bad because they encourage people with unwarranted genetic traits to breed.[40] In 1922, Margret Sanger wrote advocating for birth control, "the substitution of reason and intelligence for the blind play of instinct." Otherwise, "civilization will be faced with the ever-increasing problem of feeble-mindedness, that fertile parent of degeneracy."[41] The language is quite uncomfortable to hear at present, yet genetics do have a role in determining intelligence. Steven Pinker argued that because of the atrocities committed by the Nazis in the name of Eugenics, the academic world, based on pseudo-scientific facts, ignored the role of genetics in determining intelligence and denied the presence of racial and genetic grouping.[42] In an article named *Defending Eugenics*, Jonathan Anomaly, cited studies showing a negative correlation between IQ levels and fertility and between fertility and education. Females of high educational status in the developed world prefer to adopt children at a later stage of life rather than having their own ones earlier. Pinker argued that although this is greatly humane, it yields bad eugenic effects.[43] If part of our success in life is predetermined by our genetic makeup, would it be great if we had the ability to choose the best for our future children? wouldn't it be good for them to have less disease, less aggression, more intelligence, and empathy? No one would argue that those ends are bad at all.

Savulescu and Kahane introduced the principle of Procreative Beneficence. The principle is articulated as follows: "If couples have decided to have a child, and selection is possible, then they have a significant moral reason to select the child of the possible children they could have, whose life can be expected, in the light of the relevant available information, to go best or at least not worse than any of the others.[44] They gave an example on how a couple should delay conception for some time to avoid pregnancy during a pandemic of rubella, or how people select their partners based on predetermined criteria that carry certain advantages in their minds. For example, the level of education.[45] The authors also discussed how parents, many times, postpone having children till they secure a suitable economic situation. Based on that, they conclude that there seems to be a moral intuition to select the child who is suited to have the best possible life. The principle of BP is a maximizing principle, but it is not absolute. As long as there is no competing moral reason to argue against having the most advantaged child. The principle does not directly consider the means of having that advantaged child but offers people a chance to place other moral considerations when they are making decisions.[46] Moreover, the principle does not violate the child's autonomy, because giving him more intelligence, more empathy, and more talents and capabilities will broaden his options rather than limiting them.[47] The principle is not simply an act Utilitarian one, as it considers the virtue of beneficence and does not treat the child as a mere mean to achieve the ends of his parents.[48] That is why BP is not totally compatible with Procreative Autonomy, which gives the parents liberty to choose to select even the worst possible child. The benefit, in the end, is mainly the child's benefit.[49]

Colin Farrelly, a professor in political studies at Queen's University used Russel's definition of eugenics to formulate his argument on this subject. Bertrand Russell defined eugenic, "the attempt to improve the biological character of a breed by deliberate methods adopted to that end."[50] Farrelly stated that by applying the moral virtues of benevolence and justice, and the epistemic virtues of intellectual adaptability and humility, someone can reconsider and successfully defend eugenics.[51] For Farrelly, the eugenic practices of the twentieth century were wrong for two reasons: first, their tools were not based on a sound epistemic knowledge; for example, preventing criminals from breeding is not a scientifically proven procedure in removing bad traits from the community. Second, the forced legislation by the government violated the virtue of justice of treating people fairly, and it presumed that some people are naturally superior to each other.[52] Farrelly gave examples of a few legislations and measures, already practiced by the community, that can be considered, according to Russell's definition, eugenics in essence. Such as fluoridation of the water, vaccinations, educational system, diet, and exercise programs.[53] He argued that all those measurements tend to help virtuous polity to progress and have a better future for the coming generations. No one would debate the usefulness of those procedures and claim they are eugenics. Then he proceeded by saying: why when we discuss changing our genome, people say, "Do Not Play God !"[54] He pointed out that education, for example, does not only change the way we act in life but also changes our biology (epigenetic), and those acquired traits will be transmitted to the coming generation. For example, there are differences in our brain structure at present compared to our ancestors 50,000 years ago.(53) Farrelly cited Peter Singer's principle of preventing bad occurrences: If we can prevent a child from having a genetic disease, without sacrificing anything of comparable moral importance, then why not?[55]

Farrelly's approach was mainly focused on the procedure that a vitreous polity ought to consider and not based on what an individual might consider.[56] His approach is collective rather than personal. In addressing the objections about having less care to the disabled people if eugenics was applied, Farrelly's pointed out that a vitreous society would continue offering care to the disabled and would not judge them by their disability.[57] For example, our society at present continues to offered care to children born with spina bifida even if their mothers elected not to take folic acid during pregnancy. He cited Buchanan's argument on differentiating disabilities from the people with disability when we ought to prevent the first but still care for those who are suffering. (57)

Before moving to address the other side of the argument, it is worth mentioning some of the available procedures that can be placed under the umbrella of eugenics. We already discussed education, societal policies like vaccination, and fluoridation of water. We also discussed conventional ways like choosing partners. Another popular way, especially in USA, is Cryo-banks. Many agencies are currently advocating to offer gametes from competitive athletes, and Ivy League students to couples or single parents wishing to conceive.[58] Eggs/sperm samples have different prices according to the qualities offered, sometimes reaching thousands of dollars.[59] Another method of the new eugenics is Pre-implantation genetic diagnosis. Legislations in the UK, for example, allow couples to select for implantation embryos who do not have genetic diseases. In Australia, parents are allowed to choose the sex of their children as long as there is no family history of sex-linked disorder. (85) Compared to Amniocentesis and Chorionic villous sampling (CVS), PDG offers testing at the level of 8 cells (on day 3 of embryonic development). This is more advantageous as it does not require the abortion of the embryo if genetic diseases were to be found (of course that will depend on the parents' opinion of the moral status of the fetus, but that is a different discussion).[60]

4.4.2 A Sense of Unease

Recently, a friend of mine was diagnosed with what seems to be a psychotic disorder. The young fellow is a brilliant, cheerful, and quite a friendly college student. For many reasons, I thought that this event could somehow align with the content of those pages which I am trying to put down. In a sense, after witnessing the dramatic life changes that accompanied the diagnosis, I wondered what if my friend's parents had a chance to know about this disease predisposition and possibly prevent it? Why is there so much unease and discomfort around genetic engineering when many good things can be achieved with such highly advanced technology?

The central problem with the debate around genetic bioengineering and enhancement is directly related to its speculative nature. We do not know how the future will crystalize. What is so-called, Collingridge dilemma sheds light on this issue: it is easier to intervene and make a change when a new technology is still in its early phases, yet it is precisely that at those early phases, we lack the required knowledge to understand the dicey consequences of the technological applications.[61] Yet as George Estreich wrote, "I have few certainties, but one is that people-changing technologies are more likely to be used wisely if more people talked about them."[62]

In the arguments about eugenics and purposefully designing our future generations, people usually voice concerns not only on the goals themselves but also on the means of achieving those goals. Indeed, most parents work hard to secure a productive, healthy, and safe environment for their children, yet when it comes to changing their genome, something causing discomfort strikes them or at least many of them. I think the two cornerstone questions in this debate around new eugenics are: (1) Should genetic manipulation be equated to social engineering, education, healthy diet, and other traditional forms of parental caring? (2) Is there something called normal health and normal biology for us to cherish? Or are there just different genetic makeups resulting from a blind non guided evolutionary process? Should we accept our biology as a status quo? In other words, is there a difference between therapy (trying to restore normal health) and enhancement?

Let me first address the second question. I will later discuss the differences between therapy and enhancement in more detail, but a brief comment is warranted here. Despite the overlap between the two concepts, especially at the edges, a distinction continues to be quite valid. In medical schools presently, as far as I know, they are still teaching something called normal human physiology in contrast to pathophysiology. There is something called normal cardiac function, normal renal function, normal liver function, and normal hemostasis. There is equilibrium and harmony between organs that constitute what normal health is.[63] In the end, as Sandel's put it, "no one aspires to be a virtuoso at health."[64] Restoring a patient's cognition by treating hypothyroidism, educating a child to increase his memory and analytic capacity is different from manipulating genome to produce a creature with an IQ of 300. To treat a patient with a genetic disorder using some sort of biotechnology will not probably provoke the same unease in the collective mind of people compared to producing superhumans. To prevent a child from having psychosis in his twenties by gene editing is not similar to giving him an ageless body and an everlasting memory. What Jennifer Duonda said about treating Charles Sabine, a man suffering from Huntington's disease, "Who are we to tell him otherwise", is not the main controversy in the debate. The central issue lies in the desire to dominate, control, master human nature by advertising for eugenics and enhancement. Using a suffering patient's voice is not the ideal way to open the conversation about eugenics and human enhancement.[65]

The second issue to be addressed is whether education, providing healthy lifestyles to our or other traditional ways of care should be equated with genetic editing? Are these two approaches similar? Árnason, Vilhjálmur categorized Buchanan arguments against eugenics into three sections: Consent, Responsibility, and Instrumentalization. Buchanan believed that a child would likely accept his parents 'decision to prevent him from having an evil disease by agreeing to gene editing. However, he raised significant concerns in cases of dissonance and disagreements between the child's and his parent's wishes. He argued that in comparison to the effects of traditional ways of socialization methods, there are no counter-movements done by the child in case of gene editing. The changes in gene editing are not dynamic; they are mute. The child cannot change them nor revolt against them. In that sense, he is deprived of his autonomy to consent. He is bounded to a priori life judgment practiced upon him by his parents.[66] This takes us to the second part of the argument, which deals with responsibility. The child is no longer the lone author of his life. He had an "Alien Determination" invoked on him by his parents. That will affect his individuation in life, especially when dissonance in wishes happened.[67] Advocates for eugenics states that Nature nevertheless distributes her curses and blessings blindly; so, what is the difference? Buchanan has a profound answer "that we as humans experience our freedom in relation to something that is not totally at ours nor anyone's else disposal and control." That beginning of life in relation to unboundedness is the essence of our freedom. Our relation with Nature, or in theistic terms with God is neither totally controlled by us nor by anyone else; in other words, it transcends us.[68] In gene editing, the child becomes a tool to achieve the vision of his parents, and if gene editing became one day so powerful and profound, it might turn that child into a mere puppet.

In his book, a case against perfection, Michael Sandel describes the phenomenon of hyper-parenting. He tries to shed a beam of light on the probable anticipated effects of gene editing and enhancement culture in the light of what has been happening in the last few decades. Sandel starts by writing that parenthood is a school of humility.[69] It teaches the parents to accept a child as a gift to cherish and not as a project to master. He talks about how parents' love for their children should be unconditional one, not related to how the children look, what talents they possess, and what goals they achieve.[70] A balanced love between accepting (accepting love) the child as he is and transforming (transforming love) him in a flourishing way.[71] He argues that the problem with new eugenics lies in its desire to master and drive for perfection. He gives many examples from our contemporary time about how hyper-parenting phenomenon went so far toward pathological ends. Sandel writes about parents forcing their children into hard training in one competitive sport, which resulted in many joint-overuse injuries. The percentage of these sports injuries increased from 10 to 70% in the last two decades.[72] Educational psychologists also reports that some parents are trying to have their children diagnosed with learning disability to gain extra time for SAT preparations.[73] At present 5–7% of children under 18 in USA are diagnosed with ADHD. The prescription rate for Ritalin has increased by 1700 times in the last 15 years.[74] Another example is Katherine Cohen who offers a platinum package to help students to enter collages. The package costs parents only 32,995$.[75] These facts point toward where genetic editing could move, especially if it went beyond treatment to involving enhancement.

Sandel then moves on to discuss some practical aspects of applying new eugenics. In examining liberal eugenics, he argued that if genetic enhancement becomes equivalent to education, then the government might find herself morally obliged to offer eugenics. So liberal eugenics, in the end, do not totally reject the government's involvement in conducting the process.[76] On the other hand, concerning free market eugenics, Sandel makes a point that aligns with Estreich comment on Technological applications, being a response to what we demand and value. Sandel cites some examples of sperm and egg markets: a 50,000$ worth-Egg from Ivy League college student with SAT score of 1400. Or sperm banks with offices next to MIT and Harvard in Boston, offering donors 900$ per month for samples.[77] Of course, for more money, the costumers of those cryo-banks can get more info about the donors, like character type and analytical performance.[78] The question to be answered here is how further these phenomena could move if productive freedom continues to expand.

In his book The Abolition of Man, C. S. Lewis examined the consequences of applied psychology, birth control tools, and eugenic procedures on future generations. He argued that those who write on social matters had not learned yet to include time in their arguments as physicists and scientists do.[79] Lewis pointed out that each generation exercises power over its successors and revolts against its predecessors. Each attempt is an endeavor to control and shape the whole reality based on that generation's values. Every day that passes, the process of taking "control over" and "revolt against" becomes more successful. The future generations, each one after the other, gets weaker and more controlled. Every day we move toward extinction; many people will be controlled only by a few. Despite having more advanced technology in their hands, the future generations are weaker and not stronger because they are conditioned to live within their conditioners' value-frame.[80] Men dream to control the blind-reckless nature turned out to be the power practiced by some men over other men, with nature being only a medium and instrument.[81] This process will continue until reaching a point in which some men will gain total control on human nature, and shape how the others think, what they value, and how they will live. At that point, Lewis argued that the Abolition of Man would occur.[82] The tools that those conditioners will use to shape the rest of humans will differ in two ways. First, they are extremely more powerful than the traditional way of education and parenting. Second, those conditioners will judge their actions from outside the TAO (The common shared moral human tradition).[83] Lewis was referring to the danger of moral subjectivism in driving the conditioning process. Technologies do not progress in a vacuum. They reflect what a society value. Furthermore, even if we had a world government, without any powerful nations over others, the conduction of the conditioning process will still include the power of the majority over the minority. (79) Lewis was worried about the drive behind those procedures, a point I mentioned before in the theological section of this chapter, the necessity of the virtue of diacrisis (discernment) in guiding decisions. The internal drives that shape the process of eugenics are a key in carrying on further dialoged and discussions about this thorny subject.

4.5 Memory, Age, Intelligence and More

So far, I have tried to shed a beam of light on the different opinions regarding eugenics, and the anticipated hope and hypes of genetically modifying our future generations. However, what if a person wanted to change his own genetic makeup, practicing autonomy over his own body? Would there be any moral or non-moral issues regarding that aspiration? In the last part of this chpater, I will try to answer some of the main questions around the idea of enhancing human capacities.

4.5.1 Enhancement and Therapy

Let us begin by examining the difference between the concepts of therapy and enhancement. As I mentioned before, making a distinction between the two could be quite challenging. However, someone can grasp the difference between treating Tay-Sachs, or sickle cell disease, and improving a person's IQ level from 80 to 300?[84] The distinction in such extreme examples is obvious. However, what about improving a human IQ from 80 to 120, keeping the level within the bell shape statistical distribution of what constitutes a normal human functioning? What about improving the intellectual function of a child suffering from Dawn syndrome to become two standard deviations above the mean? Do we consider this a treatment or enhancement? In a sense, we enhanced the child way beyond many of his peers in school, yet we also treated his low cognitive capacity.[85] The technological advancement in health care reshaped and expanded the limitations in which classical medicine used to function. What about cosmetic surgery and orthodontics? Those could be considered both therapeutic and enhancing. Having good teeth alignment will prevent dental caries and also enhance the external look of a person. The same applies to many cases in which plastic surgery functions.[86] Another example to illuminate the intrinsic overlap is improving memory and vision in elderly patients. In a sense, those changes in the brain and eye functions are natural deterioration and can be framed within the picture of geriatric diseases. Treating them would be both therapeutic and enhancing. It would be great if we could slow down the aging process and preserve the vital organs' function, avoiding presbyopia and dementia.

The distinction is not quite problematic for those who subscribe to the liberal campus of genetic engineering. As Farrelly discussed, the distinction is illusory, and all that we have are different genetic makeups as a mere result of natural selection. He gave an example of obesity, arguing that those genes which helped us before surviving the lack of resources are not beneficial at present since food is quite available.[87] Farrelly also cited the Disposable Soma Theory to explain the human aging process. The theory suggests that the human genome was developed to favor the reproductive capacity rather than somatic maintenance, as survival was quite challenging in a hostile environment, and the preservation of the species was a higher priority.[88] In discussing diseases, Farrelly pointed out that there are three elements for the presence of disease: extrinsic related to pathogens (bacteria, viruses), intrinsic related to our nonperfect biology, and lastly, our lack of knowledge. He argued if a Martian visited us and offered a safe genetic solution to the aging process, and declining memory, and enhancing our suboptimal immune system, why a virtuous polity should ever refuse that help.[89]

4.5.2   Examining the Unease

The argument around human enhancement expands to involve many fields, from boosting up athletics performance, improving memory and cognitive capacity to finally reversing the aging process and the deterioration that accompanies it. Each one of those topics has its own specific concerns and particularities. My goal here is not to address all those different aspects. In contrast, I will try to focus on the deeper roots from which the stem of the concept of enhancement grows.

Bostrom argued that spending long hours in the gym might be a source of fulfillment for many people, yet others mainly aim for the ends of simply staying in shape and reversing the aging process. What if genetic editing was developed to save us the training time and gave us the results needed without needing all that hard work?[90] Someone might say that these easy shortcuts might kill the spirit of competitive sports and that we would no longer enjoy watching those athletics if we knew they are genetically modified creatures. However, as Bostrom argued, this objection is weak as those enhanced athletes might do more mind-boggling things if they are enhanced and they might charm us even more with their performances, not less. (90) What about improving a student's cognitive ability and memory by genetic editing, instead of spending hours and hours in reading and studying. Some people argue that we already drink coffee, use energy drinks to keep ourselves alert during long studying hours, so why genetic therapy is different. Why should it be a "big deal" if someone went to a commercial lab and freely paid for a cognitive boost?

**** Sandel formed his argument around the concept of appreciating the gift-edness character of our talents, the fact that we are not totally responsible for our powers, and we got them as gifts from nature, or in a religious language, from God. I commented on this idea before using Buchanan's argument about practicing our freedom in relation to something outside ours or others' disposal. Sandel argued that enhancement in sports is not only problematic because it provides an unfair shortcut. For him, our appreciation of great athletes comes from two main aspects. One is the effort they put into the game, and the other is the expression of their natural/gifted talents. He argued that the discomfort that we feel when we know someone had enhanced in competitive sports is not only related to the concept of effort and the fairness of play. We do not appreciate only efforts in sports but rather excellence.[91] If Lionel Messi only worked hard without achieving anything, no one would even knew about him. The success he achieved was essential in building his reputation. Now imagine that someone discovered that Messi was an android, and Dr. X manu-factured him. Our appreciation of his talent would fade and shifts toward Dr. X, who invented him. The more our achievements depend on the others, the less personal they become.[92]

Sandel uses this concept of giftedness to ground the anticipated consequences of genetic engineering. He argued that we would lose our humility by empowering the Promethean aspiration to master. The appreciation of the giftedness of our talents reminds us that what we own is not totally ours and keeps us humble. This appreci-ation also supports our social solidarity, in which we care for those who are unfor-tunate and unlucky to have diseases and disabilities, rather than pointing fingers at them. He made the analogy with how people pay for health insurance knowing that anyone could be a patient one day. The healthy help the sick, and the insurance money balances out. Now imagine that our genetic knowledge became so powerful in predicting our future health insurance companies got access to that info. How much our solidarity as a society would be affected? Moreover, the attempt to master our nature, as Sandel put it, would lead to "explosion of responsibility." We might reach a point that we no longer accept natural distributions of talents and the unfor-tunates, blaming each other and increasing our fellow citizens' burden for their life challenges and difficulties.[93]

Finally, I think Leon Kass has something valuable to say in this debate around enhancement. In an article named Ageless Bodies, Happy Souls, he pointed out two valuable facts about the intrinsic nature of enhancement. The first point is that with traditional ways of healing, the procedure itself (medicine), did not change the existential nature of the body but rather "served as an aid to nature's own power of self-healing."[94] C. S. Lewis made a similar point regarding traditional procedures. In the traditional way of education, the goal was to try to conform the soul to reality; the tools were knowledge, virtue, and discipline. In gene editing, the process is converted, how we can change nature, to suit our pleasures and goals. How instead of spending countless time practicing excelling a talent, we can change our code and reach the same ends.[95]

The second point has to do with the nature of the means themselves and their relationship with the ends. Kass stated that we value, discipline, and the striving to achieve, because we feel that there is something intrinsically valuable about those behaviors.[96] Many qualities and characteristics branch of the experience itself rather than from the ends of that experience. Taking a pill to reduce fear will not teach a person the meaning of courage. Drinking alcohol or taking a memory-erasing tablet will not teach a person how to cope with suffering and how to understand grief. Moreover, it is not only the nature of the means that counts but also the intelligible and conscious association between the effort and the results. Part of the Joy and meaning we experience in taking a University degree is related to our appreciation of the conscious relationship between our studying and the results of our effort.[97] The knowledge that we gain is not a one facet knowledge (one mean ending in one goal) but rather the nexus of relationships of results, that our complex nature, consciously and un-consciously gains. We not only learn from succeeding but also from failing to succeed. As Lewis put it, if "I pay you to carry me, I am not, therefore, myself a strong man."[98] I am not sure how content I will be with my knowledge if someone downloaded Harrison's Textbook in my brain. The process of learning is valuable not only instrumentally but rather intrinsically.

There are another two important points regarding the concept of enhancement. The first one was brought up briefly by Savulescu and Kahane. If a tremendous degree of control on human beings was achieved, and all people became the most intelligent, at that point, would intelligence mean anything?[99] We experience the meaning in our life by placing the phenomenon on a degree scale. Person A is more knowledgeable than B because B does not have the same degree of knowledge. However, B might be more knowledgeable than person C. The process of comparison continues, and it gives meaning to many concepts in life. I might be an excellent football player in my neighborhood, but I am no way a professional footballer. If genetic editing enabled all people to become the most intelligent, the most athletic, the tallest, the perfect musicians, what kind of a world would we be living in? The second concern is related to the complexity of the characteristics that we gain through our interactions with nature. We are oriented to increase our knowledge, but we cannot become experts in all the fields. Our life balances itself in an extremely complex way. If someone magically implanted the muscles of LeBron James on a lay person's legs, the latter might break his tibia after walking two steps. The bones are not adjusted to stand the enormous new muscle tension. Furthermore, that layperson, despite suddenly possessing the power to jump, will lack the required muscle coordination to perform fascinating slam-dunks, which takes years and years of training to form synapsis and wire circuits in his brain. How our body balances and fine-tunes our responses to reality is an extremely complex process. Enhancing a memory gene might have a similar (bone-breaking) effect on our minds.

4.6 Summary

The field of genetic engineering has advanced tremendously in the last few decades, with new discoveries and technologies opening doors to many possible future applications. Soon, we might be able to cure many hereditary diseases and reduce the suffering of many patients. Some scholars are also predicting that we will soon be able to overcome our limitations and live a better, fulfilled life by enhancing our human nature. While some of those aspirations seem to be great, they should always be guided by intellectual humility. Technology might be intrinsically value-neutral; however, its applications are usually value-guided. And to assure it has a proper application, more people need to talk about it. The worries always come from the tendency of human beings to abuse power. Now the power is more potent than ever, so the responsibility and caution should meticulously balance the drives. Nature, including ourselves, is extremely fine-tuned, and when we end up aiming at changing internally who we are, we should be very humble in acknowledging our limitations in understanding our own bodies and minds. And hopefully, the results will not be, the fading of Georgiana by Aylmer, in the lab of our life.[100]

Thought Box (IV)

When it comes to discussing genetic enhancement therapy and gene editing, there is always great enthusiasm associated with hidden discomfort, as in movies when a character tries to open the heavy wood door of a forbidden treasure-full kingdom.

Recently we had a visiting scholar at our hospital, a brilliant doctor who translated many years of her research work into a medication that changed the life of many patients with a genetic heart disease. After the lecture, we had the honor of sitting down and having a light lunch with her, the discussions were insightful, and the conversation continued for about an hour. There was a lot of passion for research potential, so much eagerness to succeed, and so much optimism to the point that it felt inappropriate to share any of my hesitations and fears.

Whether we like it or not, there is a sense of unease accompanying the conversations of genetic research and human enhancement. There is still a dread of repeating the atrocities of old eugenics, and many scientists are extremely cautious when discussing gene editing because of that history.

I don't believe the problem lies in whether gene editing is good or bad, intrinsically speaking. Gene editing and other types of research are all branches of knowledge. They are simply tools. Deontologically, there is nothing evil in correcting a mutant gene, so making a prior judgment is not helpful. Similarly, judging the matter based on the consequences is insufficient to justify or refute it ethically. And all those discussions about enhancement vs. therapy are a bit secondary to the fundamental matter at stake, namely the drives. Of course, knowledge before acting is essential, even if a person has good intentions. But the intentions behind genetic research are still the primary problem. Because of the profound power of this sort of knowledge, we should be quite vigilant moving forward. That's why you keep hearing the phrase "playing GOD" from both sides of the argument. People are afraid of each other, and

I don't blame them. Our history is full of heart-breaking atrocities as much as great acts of care and love. Deep inside, many of us yearn to have a just, peaceful, and joyful life. However, many believe that this is not attainable by human power. Therefore, there always needs to be careful supervision of research activities, especially when it comes to those types of research, such as genetic ones. After all, what matters is human-to-human relationships. As long as we can keep trust, we can all help each other reach a safe shore using all the tools God grants us. Unfortunately, that was not always our narrative, and we did not always learn the lessons well; that is why we ought to be careful.

Notes

1.   Allison, Lizabeth A. *Fundamental molecular biology*. Blackwell Pub., 2007. Page 1–2.
2.   Allison, Lizabeth A. *Fundamental molecular biology*. Blackwell Pub., 2007. Page 38.
3.   Allison, Lizabeth A. *Fundamental molecular biology*. Blackwell Pub., 2007. Page 13–17.
4.   Allison, Lizabeth A. *Fundamental molecular biology*. Blackwell Pub., 2007. Page 85–92.
5.   Wirth, Thomas, Nigel Parker, and Seppo Ylä-Herttuala. "History of gene therapy." *Gene* 525, no. 2 (2013): 162–169. Page 163.
6.   Wirth, Thomas, Nigel Parker, and Seppo Ylä-Herttuala. "History of gene therapy." *Gene* 525, no. 2 (2013): 162–169. Page 164–165.
7.   Wirth, Thomas, Nigel Parker, and Seppo Ylä-Herttuala. "History of gene therapy." *Gene* 525, no. 2 (2013): 162–169. Page 166.
8.   Goswami, Reena, Gayatri Subramanian, Liliya Silayeva, Isabelle Newkirk, Deborah Doctor, Karan Chawla, Saurabh Chattopadhyay, Dhyan Chandra, Nageswararao Chilukuri, and Venkaiah Betapudi. "Gene therapy leaves a vicious cycle." *Frontiers in oncology* 9 (2019): 297.
9.   Adli, Mazhar. "The CRISPR tool kit for genome editing and beyond." *Nature communications* 9, no. 1 (2018): 1–13. Page 2.
10.   Jiang, Fuguo, and Jennifer A. Doudna. "CRISPR–Cas9 structures and mechanisms." *Annual review of biophysics* 46 (2017): 505–529. Page 509.
11.   Doudna, Jennifer A., and Emmanuelle Charpentier. "The new frontier of genome engineering with CRISPR-Cas9." *Science* 346, no. 6213 (2014): 1258096. Page 1.
12.   Cui, Yingbo, Jiaming Xu, Minxia Cheng, Xiangke Liao, and Shaoliang Peng. "Review of CRISPR/Cas9 sgRNA design tools." *Interdisciplinary Sciences: Computational Life Sciences* 10, no. 2 (2018): 455–465.
13.   Jiang, Fuguo, and Jennifer A. Doudna. "CRISPR–Cas9 structures and mechanisms." *Annual review of biophysics* 46 (2017): 505–529. Page 506–507.
14.   Lino, Christopher A., Jason C. Harper, James P. Carney, and Jerilyn A. Timlin. "Delivering CRISPR: a review of the challenges and approaches." *Drug delivery* 25, no. 1 (2018): 1234–1257. Page 1239.
15.   Lino, Christopher A., Jason C. Harper, James P. Carney, and Jerilyn A. Timlin. "Delivering CRISPR: a review of the challenges and approaches." *Drug delivery* 25, no. 1 (2018): 1234–1257. Page 1240–1241.
16.   Wirth, Thomas, Nigel Parker, and Seppo Ylä-Herttuala. "History of gene therapy." *Gene* 525, no. 2 (2013): 162–169. Page 167–168.
17.   Sulmasy, Daniel P. "The varieties of human dignity: a logical and conceptual analysis." (2013): 937–944. Page 937.
18.   Macklin, Ruth. "Dignity is a useless concept." (2003): 1419–1420.
19.   Pinker, Steven. "The stupidity of dignity." *The new republic* 28, no. 05.2008 (2008): 28–31.

20. O'Mathúna, Dónal P. "Human dignity and the ethics of human enhancement." *Trans-Humanities Journal* 6, no. 1 (2013): 99–120. Page 104.

21. Sulmasy, Daniel P. "The varieties of human dignity: a logical and conceptual analysis." (2013): 937–944. Page 937 – Page 938.

22. O'Mathúna, Dónal P. "Human dignity and the ethics of human enhancement." *Trans-Humanities Journal* 6, no. 1 (2013): 99–120. Page 101.

23. Sulmasy, Daniel P. *The rebirth of the clinic: An introduction to spirituality in health care.* Georgetown University Press, 2006.

24. Kelly, David F., Gerard Magill, and Henk Ten Have. *Contemporary Catholic health care ethics.* Georgetown University Press, 2013. Page 10.

25. Kelly, David F., Gerard Magill, and Henk Ten Have. *Contemporary Catholic health care ethics.* Georgetown University Press, 2013. Page 12.

26. Kelly, David F., Gerard Magill, and Henk Ten Have. *Contemporary Catholic health care ethics.* Georgetown University Press, 2013. Page 11.

27. Kelly, David F., Gerard Magill, and Henk Ten Have. *Contemporary Catholic health care ethics.* Georgetown University Press, 2013. Page 14–15.

28. Nikolaos, Metropolitan, and Mesogaia Lavreotiki. "The Greek Orthodox position on the ethics of assisted reproduction." *Reproductive biomedicine online* 17 (2008): 25–33. Page 26.

29. Kelly, David F., Gerard Magill, and Henk Ten Have. *Contemporary Catholic health care ethics.* Georgetown University Press, 2013. Page 316.

30. Kelly, David F., Gerard Magill, and Henk Ten Have. *Contemporary Catholic health care ethics.* Georgetown University Press, 2013. Page 39–40.

31. Boer, Theo A. "REFLECTIONS ON ENHANCEMENT AND ENCHANTMENT A CONCLUDING ESSAY." *ETHICS AND THEOLOGY* (2012): 283. Page 288.

32. Kelly, David F., Gerard Magill, and Henk Ten Have. *Contemporary Catholic health care ethics.* Georgetown University Press, 2013. Page 313–314.

33. Kelly, David F., Gerard Magill, and Henk Ten Have. *Contemporary Catholic health care ethics.* Georgetown University Press, 2013. Page 315–316.

34. Kelly, David F., Gerard Magill, and Henk Ten Have. *Contemporary Catholic health care ethics.* Georgetown University Press, 2013. Page 41.

35. Kelly, David F., Gerard Magill, and Henk Ten Have. *Contemporary Catholic health care ethics.* Georgetown University Press, 2013. Page 86–88.

36. Balogiannis, Stavros. *Human enhancement from the orthodox point of view.* No. IKEEBOOKCH-2014-360. Aristotle University of Thessaloniki, 2013. Page 121.

37. Sandel, Michael J. *The case against perfection.* Harvard university press, 2007. Page 63.

38. Sandel, Michael J. *The case against perfection.* Harvard university press, 2007. Page 65–67.

39. Sandel, Michael J. *The case against perfection.* Harvard university press, 2007. Page 68.

40. Anomaly, Jonathan. "Defending eugenics." *Monash bioethics review* 35, no. 1–4 (2018): 24–35. Page 25.

41. Anomaly, Jonathan. "Defending eugenics." *Monash bioethics review* 35, no. 1–4 (2018): 24–35. Page 30.

42. Anomaly, Jonathan. "Defending eugenics." *Monash bioethics review* 35, no. 1–4 (2018): 24–35. Page 26.

43. Anomaly, Jonathan. "Defending eugenics." *Monash bioethics review* 35, no. 1–4 (2018): 24–35. Page 27.

44. Savulescu, Julian, and Guy Kahane. "The moral obligation to create children with the best chance of the best life." *Bioethics* 23, no. 5 (2009): 274–290. Page 274.

45. Savulescu, Julian, and Guy Kahane. "The moral obligation to create children with the best chance of the best life." *Bioethics* 23, no. 5 (2009): 274–290. Page 276–277.

46. Savulescu, Julian, and Guy Kahane. "The moral obligation to create children with the best chance of the best life." *Bioethics* 23, no. 5 (2009): 274–290. Page 278.

47. Savulescu, Julian, and Guy Kahane. "The moral obligation to create children with the best chance of the best life." *Bioethics* 23, no. 5 (2009): 274–290. Page 282.

48. Savulescu, Julian, and Guy Kahane. "The moral obligation to create children with the best
     chance of the best life." *Bioethics* 23, no. 5 (2009): 274–290. Page 283.
49. Savulescu, Julian, and Guy Kahane. "The moral obligation to create children with the best
     chance of the best life." *Bioethics* 23, no. 5 (2009): 274–290. Page 279.
50. Farrelly, Colin. *Genetic ethics: An introduction.* John Wiley & Sons, 2018. Page 27.
51. Farrelly, Colin. *Genetic ethics: An introduction.* John Wiley & Sons, 2018. Page 21.
52. Farrelly, Colin. *Genetic ethics: An introduction.* John Wiley & Sons, 2018. Page 32–33.
53. Farrelly, Colin. *Genetic ethics: An introduction.* John Wiley & Sons, 2018. Page 31 and
     page 34–35.
54. Farrelly, Colin. *Genetic ethics: An introduction.* John Wiley & Sons, 2018. Page 38.
55. Farrelly, Colin. *Genetic ethics: An introduction.* John Wiley & Sons, 2018. Page 3.
56. Farrelly, Colin. *Genetic ethics: An introduction.* John Wiley & Sons, 2018. Page 17.
57. Farrelly, Colin. *Genetic ethics: An introduction.* John Wiley & Sons, 2018. Page 100–101.
58. Bostrom, Nick, and Rebecca Roache. "Ethical issues in human enhancement." *New waves
     in applied ethics* (2008): 120–152. Page 19.
59. Sandel, Michael J. *The case against perfection.* Harvard university press, 2007. Page 72.
60. Savulescu, Julian, and Guy Kahane. "The moral obligation to create children with the best
     chance of the best life." *Bioethics* 23, no. 5 (2009): 274–290. Page 275.
61. Ferrari, Arianna, Christopher Coenen, and Armin Grunwald. "Visions and ethics in current
     discourse on human enhancement." *Nanoethics* 6, no. 3 (2012): 215–229. Page 217.
62. Estreich, George. *Fables and Futures: Biotechnology, Disability, and the Stories We Tell
     Ourselves.* MIT Press, 2019. Page XVII.
63. Sulmasy, Daniel P. "A biopsychosocial-spiritual model for the care of patients at the end of
     life." *The gerontologist* 42, no. suppl_3 (2002): 24–33.
64. Sandel, Michael J. *The case against perfection.* Harvard university press, 2007. Page 48.
65. Estreich, George. *Fables and Futures: Biotechnology, Disability, and the Stories We Tell
     Ourselves.* MIT Press, 2019. Page 18.
66. Árnason, Vilhjálmur. "From species ethics to social concerns: Habermas's critique of "lib-
     eral eugenics" evaluated." *Theoretical medicine and bioethics* 35, no. 5 (2014): 353–367.
     Page 357–358.
67. Árnason, Vilhjálmur. "From species ethics to social concerns: Habermas's critique of "lib-
     eral eugenics" evaluated." *Theoretical medicine and bioethics* 35, no. 5 (2014): 353–367.
     Page 360.
68. Árnason, Vilhjálmur. "From species ethics to social concerns: Habermas's critique of "lib-
     eral eugenics" evaluated." *Theoretical medicine and bioethics* 35, no. 5 (2014): 353–367.
     Page 361.
69. Sandel, Michael J. *The case against perfection.* Harvard university press, 2007. Page 86.
70. Sandel, Michael J. *The case against perfection.* Harvard university press, 2007. Page 45–46.
71. Sandel, Michael J. *The case against perfection.* Harvard university press, 2007. Page 49.
72. Sandel, Michael J. *The case against perfection.* Harvard university press, 2007. Page 53.
73. Sandel, Michael J. *The case against perfection.* Harvard university press, 2007. Page 55.
74. Sandel, Michael J. *The case against perfection.* Harvard university press, 2007. Page 59.
75. Sandel, Michael J. *The case against perfection.* Harvard university press, 2007. Page 55.
76. Sandel, Michael J. *The case against perfection.* Harvard university press, 2007. Page 75–79.
77. Sandel, Michael J. *The case against perfection.* Harvard university press, 2007. Page 72.
78. Sandel, Michael J. *The case against perfection.* Harvard university press, 2007. Page 74.
79. Lewis, Clive Staples. *The abolition of man.* Harper Collins, 2009. Page 56.
80. Lewis, Clive Staples. *The abolition of man.* Harper Collins, 2009. Page 57.
81. Lewis, Clive Staples. *The abolition of man.* Harper Collins, 2009. Page 55.
82. Lewis, Clive Staples. *The abolition of man.* Harper Collins, 2009. Page 64.
83. Lewis, Clive Staples. *The abolition of man.* Harper Collins, 2009. Page 59–60.
84. Kelly, David F., Gerard Magill, and Henk Ten Have. *Contemporary Catholic health care
     ethics.* Georgetown University Press, 2013. Page 308–309.

85. Bostrom, Nick, and Rebecca Roache. "Ethical issues in human enhancement." *New waves in applied ethics* (2008): 120–152. Page 2.
86. Bostrom, Nick, and Rebecca Roache. "Ethical issues in human enhancement." *New waves in applied ethics* (2008): 120–152. Page 1.
87. Farrelly, Colin. *Genetic ethics: An introduction.* John Wiley & Sons, 2018. Page 9–11.
88. Farrelly, Colin. *Genetic ethics: An introduction.* John Wiley & Sons, 2018. Page 39.
89. Farrelly, Colin. *Genetic ethics: An introduction.* John Wiley & Sons, 2018. Page 67–70.
90. Bostrom, Nick, and Rebecca Roache. "Ethical issues in human enhancement." *New waves in applied ethics* (2008): 120–152. Page 7–8.
91. Sandel, Michael J. *The case against perfection.* Harvard university press, 2007. Page 27–29.
92. Sandel, Michael J. *The case against perfection.* Harvard university press, 2007. Page 26.
93. Sandel, Michael J. *The case against perfection.* Harvard university press, 2007. Page 85–96.
94. Kass, Leon R. "Ageless bodies, happy souls: biotechnology and the pursuit of perfection." *The New Atlantis* 1 (2003): 9–28. Page 19.
95. Lewis, Clive Staples. *The abolition of man.* Harper Collins, 2009. Page 77.
96. Kass, Leon R. "Ageless bodies, happy souls: biotechnology and the pursuit of perfection." *The New Atlantis* 1 (2003): 9–28. Page 21.
97. Kass, Leon R. "Ageless bodies, happy souls: biotechnology and the pursuit of perfection." *The New Atlantis* 1 (2003): 9–28. Page 22–24.
98. Lewis, Clive Staples. *The abolition of man.* Harper Collins, 2009. Page 54.
99. Savulescu, Julian, and Guy Kahane. "The moral obligation to create children with the best chance of the best life." *Bioethics* 23, no. 5 (2009): 274–290. Page 284.
100. O'Mathúna, Dónal P. "Human dignity and the ethics of human enhancement." *Trans-Humanities Journal* 6, no. 1 (2013): 99–120. Page 115.

Bibliography

1. Adli, Mazhar. 2018. The CRISPR tool kit for genome editing and beyond. *Nature Communications* 9 (1): 1–13.
2. Allison, Lizabeth A. 2007. *Fundamental molecular biology.* Blackwell Pub.
3. Anomaly, Jonathan. 2018. Defending eugenics. *Monash Bioethics Review* 35 (1–4): 24–35.
4. Árnason, Vilhjálmur. 2014. From species ethics to social concerns: Habermas's critique of "liberal eugenics" evaluated. *Theoretical Medicine and Bioethics* 35 (5): 353–367.
5. Balogiannis, Stavros. 2013. *Human enhancement from the orthodox point of view.* No. IKEEBOOKCH-2014-360. Aristotle University of Thessaloniki.
6. Boer, Theo A. 2012. Reflections on enhancement and enchantment a concluding essay. *Ethics and Theology*, 283.
7. Bostrom, Nick, and Rebecca Roache. 2008. Ethical issues in human enhancement. *New Waves in Applied Ethics*, 120–152.
8. Cui, Yingbo, Jiaming Xu, Minxia Cheng, Xiangke Liao, and Shaoliang Peng. 2018. Review of CRISPR/Cas9 sgRNA design tools. *Interdisciplinary Sciences: Computational Life Sciences* 10(2): 455–465.
9. Doudna, Jennifer A., and Emmanuelle Charpentier. 2014. The new frontier of genome engineering with CRISPR-Cas9. *Science* 346 (6213): 1258096.
10. Estreich, George. 2019. *Fables and futures: Biotechnology, disability, and the stories we tell ourselves.* MIT Press.
11. Farrelly, Colin. 2018. *Genetic ethics: An introduction.* John Wiley & Sons.
12. Ferrari, Arianna, Christopher Coenen, and Armin Grunwald. 2012. Visions and ethics in current discourse on human enhancement. *NanoEthics* 6 (3): 215–229.
13. Goswami, Reena, Gayatri Subramanian, Liliya Silayeva, Isabelle Newkirk, Deborah Doctor, Karan Chawla, Saurabh Chattopadhyay, Dhyan Chandra, Nageswararao Chilukuri, and Venkaiah Betapudi. 2019. Gene therapy leaves a vicious cycle. *Frontiers in Oncology* 9: 297.

14. Jiang, Fuguo, and Jennifer A. Doudna. 2017. CRISPR–Cas9 structures and mechanisms. *Annual Review of Biophysics* 46: 505–529.
15. Kass, Leon R. 2003. Ageless bodies, happy souls: Biotechnology and the pursuit of perfection. *The New Atlantis* 1: 9–28.
16. Kelly, David F., Gerard Magill, and Henk Ten Have. 2013. *Contemporary Catholic health care ethics.* Georgetown University Press.
17. Lewis, Clive Staples. 2001. *The abolition of man.* Zondervan.
18. Lino, Christopher A., Jason C. Harper, James P. Carney, and Jerilyn A. Timlin. 2018. Delivering CRISPR: A review of the challenges and approaches. *Drug Delivery* 25 (1): 1234–1257.
19. Macklin, Ruth. 2003. Dignity is a useless concept, 1419–1420.
20. Nikolaos, Metropolitan, and Mesogaia Lavreotiki. 2008. The Greek Orthodox position on the ethics of assisted reproduction. *Reproductive Biomedicine Online* 17: 25–33.
21. O'Mathúna, Dónal. P. 2013. Human dignity and the ethics of human enhancement. *Trans-Humanities Journal* 6 (1): 99–120.
22. Pinker, Steven. 2008. The stupidity of dignity. In *The New Republic* 28, no. 05.2008, 28–31.
23. Sandel, Michael J. 2007. *The case against perfection.* Harvard University Press.
24. Savulescu, Julian, and Guy Kahane. 2009. The moral obligation to create children with the best chance of the best life. *Bioethics* 23 (5): 274–290.
25. Sulmasy, Daniel P. 2013. *The varieties of human dignity: A logical and conceptual analysis,* 937–944.
26. Sulmasy, Daniel P. 2006. *The rebirth of the clinic: An introduction to spirituality in health care.* Georgetown University Press.
27. Vizcarrondo, Felipe E. 2014. Human enhancement: The new eugenics. *The Linacre Quarterly* 81 (3): 239–243.
28. Wirth, Thomas, Nigel Parker, and Seppo Ylä-Herttuala. 2013. History of gene therapy. *Gene* 525 (2): 162–169.

Part III
Cardiovascular Medicine

Chapter 5
Clinical Interventions—Anthropological Approach to Advance Heart Failure and Ethical Challenges in Infective Endocarditis

Questions

- What is the difference between pain and suffering?
- Is there anything difference about heart failure patients when it comes to treating their chronic illness?
- How many times should hospitals re-operate on IVDUs and replace their infected valves? Can moral reasoning help us in finding the right answer?

5.1 Introduction

This chapter discusses two major ethical dilemmas in the field of cardiovascular medicine. First, the ethical challenges of the management of end-stage heart failure and the role of understanding chronic illnesses in the light of anthropological philosophy. Second, it examines the complex arguments regarding the futility of offering repeat valve replacement surgery for patients with IV drug abuse. Cardiovascular medicine is one of the most advanced fields in medicine, and there is an increasing body of literature regarding the ethical challenges of advanced technology. For those interested in Cardiology, like me, I hope you will find joy, excitement, and benefit in reading the following pages. For the others, who are not primarily cardiology driven, I assure you that the theoretical framework of many of the arguments and ideas presented here are not solely confined to the field of cardiovascular medicine. And you can still find practical ideas to apply to whatever patient's pathology you are dealing with.

S. Toro, *Introduction to Clinical Ethics: Perspectives from a Physician Bioethicist*, https://doi.org/10.1007/978-3-031-30804-8_5

5.2 Heart Failure

5.2.1 Scientific Review

According to the American Heart Association,[1] Heart Failure is defined as "a complex clinical syndrome that can result from any structural or functional cardiac disorder that impairs the ability of the ventricle to fill or eject blood." Based on disease progression, heart failure can be classified into four main stages. Stage A includes patients who are at risk of developing heart failure but don't have structural disease or HF symptoms, Stage B includes patients who have structural heart changes but lack any symptoms, Stage C comprises those who have both structural changes and suffer from HF symptoms, while Stage D represents patients with advanced heart failure requiring LVAD, inotropes or heart transplantation.[2]

Pathophysiologically, patients with heart failure, have either reduced or preserved ejection fraction. Uncontrolled hypertension, obesity, diabetes, hypercholesterolemia, and coronary artery disease all predispose to cardiomyopathies. Alcohol, chemotherapy, and radiation are also among the many other known causes. On the cellular level, myocardial injury leads to increase tissue fibrosis, which increases oxygen consumption and progressive ventricular dilatation and remodeling. The poor cardiac output leads to neurohormonal activation, an increase in water and sodium retention by the kidneys, which leads to a constellation of signs and symptoms, including peripheral edema, pulmonary congestion, malabsorption secondary to gut edema, cachexia, fatigue, and dyspnea.[3]

The diagnosis of heart failure is a clinical diagnosis. None of the available tests is sufficient alone to diagnose the disease. The presence of signs and symptoms usually raises concern, although most of these signs and symptoms are not specific. NT-ProBNP has excellent sensitivity, and it is the main test in the diagnostic algorithm. Echocardiogram helps with learning more about the ejection fraction, the presence of significant valvular pathologies, and wall motion abnormalities. Stress tests and cardiac catheterization give information about the status of coronary arteries and the presence or absence of coronary artery disease. Cardiac MRI has better accuracy than Echocardiogram in diagnosing infiltrative cardiomyopathy; however, high cost and limited availability are among its limits.[4]

Once the diagnosis of HF is confirmed, patients are advised to follow a strict lifestyle modifications, Such as, Reducing salt intake, monitoring their weight daily, being compliant with medications, avoiding NSAIDs, and many other things. Regular exercise is also highly recommended, and it has also been shown to increase functional capacity and improve clinical symptoms.[5] From medications standpoint, Neurohormonal blocking agents are a cornerstone of treatment, and they include ACEIs, ARBs, MRAs, and ARNIs. All have been shown to improve survival. Beta-blockers counteract the effects of the sympathetic nervous system, reduce ventricular remodeling, morbidity, and mortality. Diuretics help to decrease intravascular volume, reduce congestive symptoms, and they are essential in treating acute and

chronic presentation of heart failure.[3] Apart from Pharmacotherapies, medical technology also plays a remarkable role in the management of HF. Cardiac Resynchronization therapy is a pacing-based approach that has been shown to improve quality
of life and decrease hospitalization.[6] For patients with advanced heart failure, LVADs
can be used as a bridge to heart transplantation or as destination therapy. Evidence
is present regarding the efficacy of this technology in enhancing the quality of life
and providing a clinically meaningful survival.[7]

5.2.2 General Facts

Heart failure is one of the most prevalent chronic diseases in the United States.
In 2017, it was estimated that more than 6.5 million American adults suffer from
heart failure.[8] Among those, nearly 80% of hospitalized patients with advanced
heart failure are older than 65 years.[9] Despite the variety in incidence rate across
different studies,[10] it is expected that the prevalence will continue to steadily increase
with aging population.[11] According to Framingham Heart Study, the lifetime risk of
developing congestive heart failure was 1 in 5 for both genders in the presence of
MI, 1 in 9 for men and 1 in 6 for women in the absence of ischemic injury.[12] In 2012,
2.4% of United States population suffered from HF. In 2030, it is expected that 8
million Americans will have the disease, a quarter of these patients will be above the
age of 80.[4]

Heart failure is also a disease of a high mortality rate. After diagnosis, half of
all heart failure patients die within four years of diagnosis, and half of those with
advanced HF will die within a year of diagnosis.[13] Patients with heart failure suffer
from a significant physical and mental burden at the end of life. The prevalence of
untreated symptoms (dyspnea, pain, anxiety), poor quality of life, and depression in
the last 30 days before death exceed cancer patients.[2] The utilization of acute care
services is also higher compared to cancer patients (64% versus 39% for ED visits,
60% versus 45% for hospitalization, and 19% versus 7% for ICU stays).[14] These facts
highlight the palliative needs of this patient's population. Two Randomized Control
Trials had shown the benefits of integrating palliative care service with advanced
heart failure management. These studies revealed better outcomes with improved
quality of life, reduced morbidity, and higher spiritual well-being.[15] Despite the
evidence, studies also have shown that in heart failure patients, palliative care and
hospice services are still underutilized at the end of life.[16] Researchers suggested
many reasons. Some of them are related to the health care system and lack of communication between physicians and patients, while others were directly related to the
disease nature and trajectory of Heart failure.[17]

Besides being a disease with high mortality, heart failure is also very costly. In
2015, the annual cost of HF in the USA was estimated to be 44.6 billion dollars. By
2030 the annual cost will likely reach 97 billion dollars.[18] The high financial burden of
managing chronic diseases, including heart failure, raises significant concerns about
distributive justice. While many patients would prefer to receive advanced, costly

therapies, policymakers and society might view these treatments as unnecessary. In the end, medical resources are limited and have to be used judiciously; when it is right to use or not to use certain technologies and treatments is ethically very challenging.[10]

Lastly, the advancements in medical technologies have complicated the management at the end of life. Many advanced heart failure patients are currently living with implanted defibrillators and LVADs. Due to the paucity of organs, many patients are now receiving LVADS as destination therapy.[19] Studies have shown that LVADs improve patients' survival and quality of life.[20] However, the use of these devices was not free of complications. It is well known that they are associated with significant morbidity, including infections, strokes, and device failure. In light of the increasing availability and elevated cost of these technologies, many questions are raised on who should receive them. Also, how should physicians handle device de-activation? Should advance directives be discussed at the time of implantation? What are the roles of palliative care? All these questions require careful examination to come to a reasonable consensus on how to deal with these challenges.

5.3 Understanding Chronic Illnesses

5.3.1 Person and Value

Cassell defines a person "as an embodied, purposeful, thinking, feeling, emotional, reflective, rational human individual always in action, response to meanings and whose life in all spheres points both outward and inward."[21] To illustrate more on the value of treating patients as unique individuals, I shall examine two examples from well-known characters in literature. The aim is to demonstrate the different dimensions that shape our understanding of human beings. This analysis will help us to understand the multiple levels of meaning of persons in general and consequently, patients.

Raskolnikov, the main character in Dostoevsky's novel Crime and Punishment, was described in the narrative as an isolated law student, poorly dressed, slim, well built with dark eyes and brown hair.[22] Raskolnikov believed that the world was divided between ordinary and extraordinary people. Ordinary people lived in submission and did not have a right to break the law. On the contrary, extraordinary people had the right to rule, transgress, and were allowed in a utilitarian sense to achieve goals even if that led to committing crimes.[23] Don Quixote is another example of a very well-known character. The old gentleman from the village of La Mancha, who had wrinkled features and was in his fifties. The old daydreaming knight valued chivalry stories so much to a point that he ended up selling many acres of his land to buy books about knights.[24]

If the main goal of medicine is to relieve suffering, and if suffering cannot be understood solely through empirical science, then medicine cannot achieve its goals purely by scientific methods.[25] As I will be discussing later in this chapter, suffering is a personal and complex experience of an individual in a given society. To help a sufferer, the physician must first learn how to deeply understand his patients. This process of learning about humans (patients) is not as simple and easy as it seems, especially in the medical world. Let me illustrate more on this point. Every individual presents himself and can be viewed by others in three different dimensions.[26] The first dimension is *Empirical*, revealing specific facts about that person. We mainly grasp those facts by the function of our senses. The second is *Moral* or, in a more general sense, value-based. For example, what this person likes, hates, believes, cherishes, and many other judgments that define how he interacts with other humans. The third is Aesthetic, the harmonious presentation of a person in his environment. This level or dimension interacts on one side with the previous two dimensions and on the other side with the surrounding society. Now, let us move back to our previously discussed characters and try to apply this analysis. Raskolnikov was well-built and had dark eyes, Don Quixote had a wizened face, and was fifty-year-old. These are empirical facts about the two characters. Anyone reading the text can imagine the shape of dark eyes and wrinkled faces. When someone presents with a maculopapular rash, fever, jaundice, swollen joints, those are all empirical facts about patients. Interestingly, they are not only objective but also subjective. The degree of yellow color that two physicians observe might not be the same, yet both will agree that the patient is jaundiced.

The second dimension in understanding persons is value-laden.[27] We all behave based on our beliefs. We have a set of moral principles that guide us in determining what is right or wrong. For example, Raskolnikov believed himself to be an extraordinary man. His opinions and resentment led him to justify killing Alyona Ivanova; on one side, to get money and help his family, and on the other side, because he believed she was an awful deceitful human and killing her would offer a utilitarian benefit to the society. Those values that he held guided his behaviors and actions. However, values are not confined only to the moral realm, and they can also be perceived outside the domain of morality. Humans value different things in life, like soccer, music, certain restaurants, and cuisines; none of them has any relation to moral judgments. If science is intrinsically value-neutral, then in order for medicine to achieve its goals, it has to engage another agency to patients' care. Undoubtedly, values are always present and impactful in the clinical encounter. On one side, the patient has a set of values that he brings to the examination room. On the other side, the caring physician also practices based on a system of ethics and beliefs. For example, a physician might or might not agree with Euthanasia or abortion; a patient might accept or refuse blood transfusions or vaccinations.

The values also have a societal dimension, forming a wholeness. They harmonize with what the culture and community consider to be a moral priority—for example, caring for disabled people and children or caring for the wellness of the environment. In order to help suffering patients, physicians must actively learn their patients' values. Although personal themes like compassion, prudence, courage, agape, cannot be objectively measured, they all have a specific resonance in the collective mind of people. When we read that Don Quixote was idealistic and honest, we all have a shared sense of what being ideal and honest means.

Lastly, Values also have Aesthetic dimension; they do not manifest outside the natural order of someone's persona. Let me elaborate more on this point. Many friends, family members, and people we know behave in a specific way according to their values. How someone dresses, talks, treat others, are all reflected harmonically in his character. When this harmony vanishes or a known friend to us behave extraordinarily—outside his persona—we all feel that something is not right. The way Raskolnikov and Don Quixote were presented in the novel, serve the purpose of the characters in a harmonious way with the surrounding environment. After a year of medical training, many resident doctors develop this clinical sense of identifying that something is "fishy" and not right in a patient's presentation. Whether it is family dynamics, secondary gain, or, as commonly referred, "something just does not look right." All these judgments are value judgments that include a sense of differentiation of what is felt not to fit in the harmonious, orderly picture.[28]

5.3.2   Suffering in Chronic Illness

Before discussing how patients suffer from chronic illnesses, it is essential to emphasize a few key points. First, the concept of illness is broader than the concept of disease. An Illness is a multi-level manifestation of a disease. It reflects the changes that inflict a human being from the molecular to the social and spiritual level.[29] Second, besides the common symptoms and shared features of different diseases, illnesses are unique phenomena that present differently between different patients. Third, when we talk about acute illness, most of the time, the treatment decision is mainly determined by the physician, for example, treating pneumonia or myocardial infarction. On the other hand, in chronic illnesses, the treatment, most of the time, involves shared decisions and opinions between the doctor, the patient, and sometimes family members.[30] In chronic illness, the disease is not sensed by the patient as a separate inflicting entity; it is now a part of him, an ongoing process, affecting his daily thoughts, activities, and interactions with the others.

Suffering can be viewed as a conflict on two main levels; the first is an intrapersonal conflict between the patient's mind and body; the second is an interpersonal conflict between the patient and society. It is quite common in medical practice to encounter patients who stopped taking care of their bodies. Some of them, for example, are morbidly obese who got used to urinating in bed. Some stop shaving and cutting their nails; others ignore a foul-smelling foot ulcer for months, and many do not shower for weeks. The question to be asked here is, why? Why do these patients behave in this way?

The human characteristics of behavior per Arthur Lovejoy involve three components: Self-esteem, Approbativeness (desire to be approved), and Emulativeness (desire to be superior). The chronically ill person is incapable of fulfilling these characteristics. Cassell gives a powerful example on a street graffito: THE WOLRD

WILL KNOW OF ME; he commented, people who walk with crutches do not write such slogans. The self-esteem of a lady after a bilateral mastectomy, the desire to be superior in a patient with bad multiple sclerosis, the Approbativeness of a patient with multiple disfiguring burns, are all a few examples of how a chronic illness presents an inner conflict. Many of these patients end up having a sense that their bodies had failed them; the body is now the enemy who is standing against the teleological flourishing and fulfillment of desires.[31]

On the other side, the intrapersonal conflict is a conflict between the patient and the surrounding community. After perceiving himself as a sick person, the patient now has to face life challenges. To try to catch up with the requirements of daily living and with the customs commonly shared between people. To climb stairs, get on and off a bus, sit on a toilet, shake hands, and many other countless common daily behaviors. The confidence is now replaced by low self-esteem, the strength by vulnerability, and the sense of usefulness by uselessness.[32] Patients start to feel that they are not only a burden on themselves but also a burden on their beloved ones and their society. Consequently, feelings of anger, anxiety, and depression lead many of them to isolation. They give up, stop taking medications and stop being compliant with the physicians' recommendations; as a result, more disease exacerbations occur, leading to more hospitalizations, less freedom, and consequently, patients get trapped in this vicious cycle.

5.4 Heart Failure, Anthropological Analysis

So far, I have tried to shed light on some of the abstract aspects of patients' suffering. Yet despite being shared as a fact in our existence, the experience per se always manifests uniquely in each individual sufferer. In the following passages, I will attempt to ground the illness of heart failure on philosophical anthropology. This approach-despite being to some extent incompatible with the way medicine is practiced at present- is quite essential to understand the nature of the illness and its complicated relationship to each patient within and without.

Before start examining Heart Failure, let me provide a brief introduction to this anthropological approach. Contemporary medicine is accused of limiting its attention mostly to the finitude of humans.[33] The holistic approach to alleviating human suffering is quite challenging. Etymologically, the word *heal* means to make a whole, to take a holistic approach to treat the sufferer. As the Jewish philosopher Abraham Heschel said, "To heal a person, one must first be a person."[34] The understanding of humankind in philosophical anthropology proposes that each individual is essentially a spiritual being in a relationship. *Being* is a relationship, and to know anything necessitates an understanding of a nexus of relationships. Matter on the subatomic level is a relationship. What constitutes matter is not a combination of unimaginably tiny particles but a set of temporary dynamic relationships of electromagnetic fields.[35] Anything, whether being a person, a concept, a bacterium, or a disease, is a form of a relationship.

Illness can be understood anthropologically as a disturbed relationship.[36] Our ancestors, throughout medical history, wisely grasped a hint of that reality. The Humoral Theory of Illnesses—despite its inaccurate presumptions—is a good example of this notion. An understanding of diseases as a disrupted relationship between the four main body humors: blood, phlegm, Yellow bile, and Black bile.[37] Another interesting example in medical history is the Shaman, the witch doctor who tried to heal by restoring the relationship between the patient and the Gods.[38]

The disrupted relationship that leads to illness can be viewed on four main levels: two intra-personal and two extra-personal. Biologically, a disease is a disturbed relation between different organs. For example, diabetes is an unbalanced relationship between the pancreas, intestines, the adipose tissue, and the liver. This change leads to chronic elevation of blood glucose, which eventually causes many micro and macrovascular complications. C.diff Colitis is another good example. Similarly, it results from a change in the balance of the natural microbiome in the colon.[39] A change in the normal order. Ergo, a disease is not a bad body inside a normal body, but rather a disturbed hemostatic state, a disorder, and chaos replacing harmony within and between various organs.

The second intrapersonal level is a Mind-Body conflict. The illness manifests with psychological symptoms: fear, isolation, shame, and many other complex emotions that strike the patient and lead to significant anxiety and depression. As I mentioned before, the patient might feel that his body had failed him. The body is perceived as a separate entity standing against someone's existential desires and life fulfillment.

On the other side, the extra-personal levels of illnesses are social and spiritual.[40] When illness strikes a patient, it does not only disrupt the relationship inside his body but also affects his relationships with the surrounding community. Chronic diseases muddle family dynamics, workplaces, and other social circuits around the patient. It affects his daily interactions and communications with others. A few random comments from cancer patients can help us apprehend this point: "people can tell that something is wrong when you are bald; they don't want to be around you because it is too scary!" Another patient said, "It is really pity; they are trying to tell me how sick I am" commenting on starting to work only part-time. Many other examples of similar observations in our daily lives are quite common.[41] The effects of such dramatic existential changes in a patient's life are part of the illness, a part of reality, and an important aspect that each physician has to be so careful and prudent in addressing.

Lastly, the Spiritual dimension is the final level of relationship in understanding illnesses. It has to deal with someone's own searching for transcendent meanings. Later on, I will discuss in detail the difference between spirituality and religion, but for now, let me try to give a brief preface. Every person can be considered essentially a spiritual being. In a sense, an object always transcends its meaning. Each time we try to learn about something, the phenomenon keeps transcending.[42] Someone can describe horses in 2 sentences, but he might spend his whole life writing about horses without closing a book chapter. Additionally, as we move further in our attempt to understand anything, the knowledge becomes more immaterial, metaphysical, dealing with values, emotions, and aesthetics. Throughout history, this relation between humans and the transcendence manifested widely in religious practices, cultures, traditions, art, and philosophy. For example, when the black death struck Europe in the fourteenth century, it raised existential questions about the meaning of life and suffering. People responded by reasserting religious beliefs and emphasizing certain moral values during the crisis.[43] Chronic illness always triggers the patient to ask, Why Me, and Why Now? Both questions are, without doubt, transcendental questions.[44]

5.4.1 Intrapersonal Relationship

Heart Failure is a systemic disorder caused by the inability of the heart to provide sufficient cardiac output and to accommodate volume preload to maintain metabolic demands.[45] Regardless of the underlying etiology, HF affects almost every single organ in the human body in a bidirectional manner. The purpose here is not to provide a detailed pathophysiological explanation of multiorgan dysfunction associated with HF, but to re-emphasize understanding the disease as a disturbed relationship. Cardiorenal Syndrome is a good example to start. A complex syndrome with five different phenotypes reflecting the extreme connection between the heart and the kidneys. The prevalence of renal dysfunction in CHF is reported to be almost 25%.[46] In CRS the failing heart, whether acutely or chronically, leads to renal damage. A cascade of hemodynamic and neuroendocrine responses occurs by activating RAAS (Renin Angiotensin Aldosterone System), increasing TNF-Alpha, IL-1, and IL-6 leading to Uremic Cardiomyopathy, direct cardio depressant effects, and progressive fibrosis.[47]

The Gastrointestinal system is also among the major systems affected in CHF. Increased gut edema secondary to volume overload and decreased perfusion both lead to an increase in gut permeability and bacterial translocation. This translocation of gut flora was linked to increased levels of blood TNF-Alpha and chronic inflammatory state associated with cardiac cachexia.[48] Anemia is also quite common in CHF patients, usually in the form of anemia of chronic disease. The inflammatory state caused by decreased tissue perfusion leads to relative bone marrow suppression, decrease reticulocyte count, and Ferroprotein (Iron gets trapped in Macrophages). Anemia also reversibly affects the cardiovascular system, causing a hyperdynamic state of circulation due to decrease blood viscosity, leading to increase preload (Frank-Starling mechanism), Left ventricular hypertrophy, and probably proatherogenic role through changes in the blood flow pattern.[49]

The psychological effects of HF on suffering patients form the second level of disturbed relationship, namely the conflict between the patient's mind and body. Anxiety and depression are both quite common in HF population. A meta-analysis of 36 studies revealed that one-third of HF patients reported depressive symptoms, and 19% met the criteria for depressive disorder.[50] Anxiety also is highly prevalent, and is present in about 30% of HF population.[51] In terms of outcomes, major depression is a known poor prognostic factor in HF population that has been linked to frequent hospitalization and mortality.[52] The presence of concomitant anxiety is also known to further worsen morbidity and mortality.[53]

Regarding the treatment of depression in CHF patients, cognitive behavioral therapy (CBT) has been shown to be effective in this patient population. In a randomized controlled trial of one hundred fifty-eight patients, depression scores were lower on CBT arm compared to usual care.[54] On the other hand, the benefit of pharmacotherapy is less clear. Two RCTs showed no significant difference in reducing depression scores with SSRIs compared to placebo.[55] In SADHART- CHF trial, there was no significant reduction in HDRS total score between Sertraline and Placebo groups [The mean change between groups was -0.4 (95% CI -1.7 to 0.92; P $=$ 0.89)]. Similarly, in MOOD-HF trail, there was so statistically significant difference in MADRS (Montgomery-Asberg Depression Rating Scale) between the two groups after treatment (Escitalopram vs. Placebo) [Difference in MADRS was—0.9 (95% CI, -2.3 to 0.4; P $=$ 0.4)]. The authors concluded that the underlying pathophysiology of depression in HF patients might be different, as most of the clinical trials that evaluated the use of SSRI in depression had not included patients with somatic diseases. In both trials, the treatment of depression with Sertraline and Escitalopram did not reduce all-cause mortality.

5.4.2   Extra Personal Relationship

The third level of relationship in the anthropological analysis of heart failure is Sociological. Bury describes the chronic illness as a 'biographical disruption' of daily life, affecting, on one side, the individual sufferer and, on the other side, his social network.[56] Social support is defined as 'harmonious social interaction.' It consists of four levels: (a) socially, the effective integration in a culture; (b) cognitively, the ability to intellectually communicate with others; (c) emotionally, it connotes a feeling of security, (4) physiologically, it leads to less anxiety and vulnerability with decreased levels of cortisol, epinephrine, and other related stress hormones.[57] In terms of outcomes, a cohort study of 292 patients with HF showed that lack of social support was an independent risk factor for fatal and non-fatal outcomes in this patient's population.[58] Another cohort study showed that single marital status, which might be an indicator of lack of social support, was an independent risk factor for hospital readmission and mortality (adjusted HR 2.1, 95% confidence interval 1.3 to 3.3).[59]

Lastly, the coping of family members of HF patients varied between positive and negative attitudes. Some reported feelings of frustration, resentment, and suffocation, given the high demands of taking care of these patients. Uncertainty and unpredictability of the disease course were major contributors to the presence of such feelings.[60] On the other side, many family members of HF patients reported feelings of satisfaction, confidence, and reward by being able to take care of their loved ones.[61] Thus a holistic approach is indicated to address the needs of caregivers, to offer them psychosocial support, and to help them maintain a sense of normalcy.[62]

The connection with the transcendent forms the last level of relationship in the anthropological context of chronic diseases. The points I am trying to discuss here might not be confined solely to heart failure per se. However, it is quite pivotal for any practicing physician to appreciate the spiritual dimension of a suffering experience. First, let me start by discussing the difference between Spirituality and Religion. Per Sulmasy, spirituality is the characteristics of the qualities of someone's relationship with the transcendent, regardless of whether this transcendent is called God, or something else.[63] People always seek to comprehend this transcendence, that is not totally inside them, not completely outside them, and neither equal to them.[64] As mentioned before, each meaning transcends its object. Even an atheist who rejects the excising of such transcendent has already established his relationship in a context of rejection. Religion on the side is a specific creed of certain beliefs about God that is shared by a particular community and usually involves a tradition, language, and texts.[65]

Suffering in the spiritual context can be viewed as a result of human finitude. Humans are oriented toward the infinite, yet they are aware of their finitude. Perhaps this agonizing paradox is the greatest mystery of humankind, and it broadly manifests in intellect, moral, and aesthetic facets.[66] Every man is intellectually oriented to learn more and expand his knowledge, yet deep in his heart, he knows that ultimate knowledge about anything is not achievable. Humans are also blessed with the judgment of good and evil. Though they strive for goodness, they find themselves living in a world drowned in evil and full of malevolence and injustice. Additionally, people are oriented toward beauty, yet they know, and quite well, that they will grow old and become sick, and their beauty will gradually fade down.[67] So to speak, they always feel a sense of limitedness in a search for the ultimate. Before moving to the next idea, it is important to make a distinction between pain and suffering. Not every pain leads to suffering, and human suffering is not solely caused by pain. Pain is a physiological phenomenon. A patient's experience of pain will turn into suffering when it contributes to his overall experience of limitedness and finitude. The experience of suffering is broader than the experience of pain.[68]

When a chronic disease inflicts a patient, it strikes him with the reality of his finitude. It raises questions about his values, his relationships, and his life's meaning. During our daily medical practice, these three notions are usually covered under other terminologies. The question of someone's values is placed in the category of *dignity*. The concept of meaning usually refers to our discussions about *hope*. Lastly, the word *closure* usually refers to restoring relationships and reconciliation.[69]

In the face of his limitedness, a sufferer of chronic disease asks many existential questions. For example, does my life have any value? Can I still have any worth even when I am unproductive? Both of these questions are questions about dignity. First, a distinction has to be made between attributed dignity and intrinsic dignity. Intrinsic dignity is the dignity that every human being possesses because he is simply a member of this human family. On the other side, attributed dignity is the sense of value that someone has based on his achievements, social status, and other forms of power degree.[70] As health care professionals, we should always try to remind our patients that they have an indelible intrinsic value. That no matter how limited and dependent they become, they still have priceless value. By bathing, feeding, and empathizing with them, we are contributing to relieving part of their suffering. We are reminding these patients that despite everything that happened, they are still intrinsically valuable.[71]

The second question that usually arises in the context of chronic suffering is a question of connectedness. The reconciliation and restoring of a broken relationship. On one side, between the patient and his beloved ones, and on the other side, between him and God. This reconciliation requires a person to accept his own imperfections and to ask forgiveness.[72] In the paradox of searching for ultimate love and the presence of suffering as a reality, every patient has to be reminded that he is a beloved human being, and all the possible efforts have to be made to help him connect with those whom he loved.[73] Within this context finally lies the last question to be asked, the question of Hope. Hope not in the sense of optimism where things might end up being well, but rather the Hope that makes sense no matter how things turn out to be; the Hope that the present events will be reinterpreted, and all questions will be answered.[74] In Christian Theology, Jesus was the physician who restored the broken relationship between humankind and God "your sins are forgiven" (MT 9:1–8). By his miracles of healings, he also restored the relationships inside families (Lk 7:11–17).[75] For a suffering patient, to experience the illness and feeling invaluable is an ultimate indignity, to suffer without believing in higher meanings is ultimate hopelessness, and to suffer alone is absolute alienation. Health care professionals are invited to transcend the meaning of suffering through compassion and love.[76]

The body of literature discussing the spiritual needs of heart failure patients supports the points mentioned above. Evidence suggests a high degree of spiritual distress in the heart failure population. Furthermore, spiritual well-being has been associated with lesser degrees of anxiety/depression and better quality of life.[77] Most of the studies examining the spiritual needs of heart failure patients involved subjects with advanced disease (NYHA III/IV), and less evidence is available on milder stages.[78] Murray and his colleagues showed that the spiritual well-being of heart failure patients gradually declined throughout the disease course, with some fluctuations paralleling the physical well-being trajectory.[79] Searching for connectedness, value, and meaning was a predominant theme. Patients struggled to maintain their self-image and dignity as the disease progressed, and their physical ability declined.[80] A recent review of spiritual coping in advanced heart failure patients affirmed these points. Patients sought to find meaning and hope through their illnesses. Some of them found solace in their faith in God, while others had more negative emotions of regret and resentment. Many patients searched for reconciliation and found comfort in reconnecting with their friends, family, and beloved ones.[81]

In contemporary times, adopting a holistic approach to patients' care is quite challenging for many reasons. With health care being transformed into a commercial industry, the classical virtues of medicine like compassion, fidelity, and fortitude are quite challenging to apply. The main theme of the health care industry is efficiency. With clinical encounter time being restricted to a few minutes, it is unthinkable to talk about reconciliation and building relationships with patients. Indeed, talking about transcendent values in such a limited time is even more challenging. Medical practice at present is mostly confined to the first biological level of relationship, enhanced by the adoption of scientific reductionism and shielded by evidence-based medicine.[82] To act for the good of a patient, physicians must learn to read the narrative of a patient's illness. Even if the clinician—and this is likely the case—is unable to solely handle all these aspects of care, being aware of them is a first good step. Following that, he can introduce, guide, and refer patients to the widely available resources. Only then, true healing might be possible.

5.5   The Dilemma of Valve Replacement Surgery in IVDUs

Infective endocarditis is a bacterial infection of the endothelium of the heart.[83] It is a disease of poor prognosis.[84] Studies have shown a mortality rate between 11% and 25%, while the estimated five- year mortality rate reaches up to 50%.[85] At the beginning of the 21st century, infective endocarditis (IE) cases secondary to IVDU represented only a small percentage of overall IE cases. However, this percentage has significantly changed over the past few decades. One study reported an increase in the proportion of IVDU related endocarditis from 14.4% in 2002 to 26% in 2014.[86] Another study reported a 5.1% increase in IVDU-IE related hospitalization between 2000 and 2013.[87]

Since 2005, there has been a shift in the recommendations of professional medical societies toward earlier surgical interventions. The 2015 AHA guidelines recommended early surgical treatment in the presence of high-risk features such as high virulent organisms, persistent bacteremia, complete heart block, and valvular dysfunction leading to HF symptoms.[88] In terms of surgical outcomes, data regarding this issue is conflicting, with some studies reporting higher mortality rates among IVDUs compared to Non-IVDUs,[89] while others show no difference or quite the opposite outcomes with lower mortality rates.[90] In fact, a recent Meta-Analysis of thirteen studies (N = 1593 patients) showed no statistically significant difference in 30-day mortality between IE in IVDUs and Non-IVDUs. Interestingly, the RR was 0.88, suggesting probably better outcomes in IVDUs. However, the same study did not evaluate long term outcomes and mortality due to data heterogenicity.[91] Nevertheless, it is well known that relapse to drug use and reinfection are quite common. In a retrospective study of 102 patients in a large academic center in Boston between 2004 and 2014, 50 patients got readmitted, 28 of them had already relapsed to drug abuse, and 14 had valve reinfection (Median time to hospital readmission was 216 days).[92] Another study of 109 patient admitted to chemical dependency service reported a 59% relapse rate within a week of hospitalization discharge.[93] In light of this evidence, many patients present with recurrent valve infections. Surgical and medical teams usually face medical and ethical challenges of offering these patients another valve replacement surgery. In the following passages, an argument supporting each side of this ethical dilemma will be provided, discussing the different views regarding the concepts of futility, individual responsibility, medical and social justice.

5.5.1 Surgery Should not Be Offered

Some scholars suggested that redo-valve replacement should not be offered to those who return to use IV drugs. The argument is generally made on three main levels. The first is personal responsibility. Advocates for this position assert that people should be held accountable for their life decisions—as long as patients have autonomy and capacity to understand the consequences of their actions, and as long as no significant social pressures are influencing their behaviors.[94] Although environmental and genetic factors partially play a role in addiction disorders, some people do succeed in getting clean. Moreover, on a juridical level, the government punishes criminals who act under the influence of drugs, clearly confirming their personal responsibility for their actions. Similarly, physicians should also be responsible. As professional members practicing in a health care system, they are entitled to assess the clinical presentation and to decide whether the treatment is futile.[95] If the patient was acknowledged before discharge about the risks of re-infecting the new valve by continuing drug abuse then presented with prosthetic valve infection, it was the result of his/her failure to comply with the plan that leads to those consequences. The physician has the right to decide in such cases if subsequent surgeries are futile in light of a patient's recidivism.

The second argument is framed around the safety of the staff. By deciding to proceed with surgery, the surgeon is putting the wellbeing of himself and the other team members under many risks. IVDUs are a group of patients with a high prevalence of contagious blood born viruses like HIV and HCV.[96] The risks of transmitting viral infections, both intra and postoperatively, are real.[97] Several studies reported the risks of acquiring HIV and HCV after a single exposure to be 0.5% and 1–2% respectively. Furthermore, cardiac surgeries have the highest rate of mucocutaneous exposure compared to other surgical operations reaching up to 50%.[98]

Lastly, on a societal level, physicians and other health care professionals have to act as good stewards. They should be responsible for using precious medical resources judiciously. In the end, money, time, and human resources are limited, and society cannot continue offering care to irresponsible people. (94) Although someone might argue that we do not refuse to offer diabetes treatments and cardiac stents to people who continue to overeat and smoke, the analogy is not quite synonymous. Although smoking and overeating are bad traits, they mostly affect the individual sufferer who smokes and overeats. Smoking, for example, does not lead to the same level of societal damage as IV drugs. Smoking does not cause family breaks, unemployment, and elevated crime rates. Moreover, drug abusers usually rely on community help to survive and are unable to contribute functionally and financially to the society in which they live.

5.5.2 Surgery Should Be Offered

Generally speaking, there are three counterarguments regarding this ethical dilemma. The first is the *personal bias argument*.[99] Patients' failure to comply with treatment plans usually triggers significant frustration between health care professionals. At the same time, it is well known that people have difficult times changing their personal habits.[100] Indeed, in the case of IV drug use, with its known physiological and psychological effects, changing lifestyle and breaking the addiction is unimaginably more challenging.[101] The frustration with recidivism must not affect the medical decisions of health care professionals. In the end, illnesses are socially crystallized concepts. Even if someone argues that a heroin user was responsible for his life decisions, we do not think in the same way about a motorcyclist who presents twice to the ED with many fractures and life-threatening trauma. Both those two individuals made risky decisions in their lives. No one would dare to consider not offering an Olympic skier a femur or neck surgery because he did that to himself by jumping 350 feet with a speed of 65 miles/hour.[102] The social stigma of IVDUs as a patient population should not affect the clinical judgment of a physician at bedside.

The second argument is a *Futility argument*. Should we consider redo-valve replacement surgery in recidivist patients a futile treatment? To answer this question, we should first review the different definitions of the concept of Futility. Since ancient Greek times, the duty of physicians to not give a useless treatment was recommended and probably practiced. Plato wrote, "For people whose lives are always in a state of inner sickness, Asclepius did not attempt to prescribe a regimen to make their life a prolonged misery."[103] According to the Oxford Dictionary, futile means 'Leaky, Vain, Failing of the desired end through intrinsic defect.'[104] This articulated concept entails three notions: A goal, an action to achieve a goal and a probability or a degree of certainty that the procedure will fail to achieve this goal. In literature review, we can find that some scholars tried to define Futility in Quantitative and Qualitative terms.[105] They suggested that for a treatment to be futile, it should fail to produce the desired effect in 100 cases. The justification of the cutoff was based on the same statistical evaluation of clinical trials, where a P value of 0.001 is considered sufficient to reject the null hypothesis.[106] Moreover, they added that a Qualitative component in assessing medical futility is also necessary. The applied treatment or procedure should not only produce a physiological effect but also results in an acceptable quality of life. For example, using artificial nutrition in PVS is considered a futile treatment in the light of overall patient's clinical condition and prognosis. Other scholars took a more restrictive approach in defining futility, confining it only to those treatments that should not be used because they are against the standards of care. For example, treating pneumonia with dialysis or viral infections with antibiotics. The decision, by far, is a medical decision and not an ethical one.[107] Historically, the concept of futility was initially introduced to discuss the ethical debates around life-sustaining treatment. It was later generalized due to ideological (the emphasis on the concept of autonomy) and economic factors (costly treatments) to involve various treatments and procedures.[108] Now, considering what we have already discussed, let us return to our case. Is valve replacement surgery a futile treatment? An action that would not achieve the desired effect no matter how often it was repeated? The answer is probably No. Valve replacement is a successful treatment for infective endocarditis. Whether or not the patient would relapse to drug abuse and re-infect his new valve is a non-grounded assumption. Yes, the odds of relapse are high, but some patients do succeed in breaking the addiction cycle and staying clean.

The last argument is the social justice argument. Should physicians take responsibility for allocating limited medical resources? To answer this question, we should first differentiate between social and medical justice. From a Medical standpoint, physicians should not refuse to give a patient a beneficial treatment based on social considerations, whether the sufferer is a prime minister or a criminal, as long as the medication is useful in treating the underlying disease. The process of Rationing resources should not be made at the bedside.[109] First, physicians are not trained to be social economists to approach such a complex issue. Second, any sort of such rationing will violate the fidelity of trust defining the patient-physician relationship.[110] Although social justice, in essence, is a comparative concept (by giving someone a treatment, we are holding the treatment from another person), the decision regarding whether and how many times an IV drug user deserves getting another valve from an economic standpoint should be made only at societal levels, and not at the ICU beds. Moreover, social justice raises two essential questions: first, should we consider IV drug users full members of the society and should we offer them all the rights and privileges, despite their recidivism? Second, in light of limited resources, would the society take a utilitarian vision of allocating resources between people for the benefit of the majority? Or it will decide to approach this problem from an egalitarian perspective, giving every individual equal access to the best possible equality of opportunity. Clearly, such a complex issue has no room for discussion at the bedside.[111]

5.6 Summary

This chapter examined the phenomenon of heart failure as a chronic illness. It discussed the various levels of existential changes that inflict a human being when reality labels him "sick." Understanding the multiple needs of our patients is pivotal for offering holistic care. Unfortunately, confining the practice of medicine only to the biological level has resulted in a plentitude of pathological manifestations. Heart failure is one of those chronic illnesses that we need to address individually. We all need to learn how to read our patients 'narratives, we all need to learn how to understand what our patients value, and how to help them transcend their suffering into higher meanings of reconciliation and hope. Perhaps, only then, would true healing be possible.

Finally, regarding infective endocarditis in IVDUs, I tried to summarize the main arguments supporting each side of this ethical debate. Clearly, the dilemma is very complex, and the answers are not solely confined to the field of medicine. The primary take points are the following: (1) Personal responsibility alone is not a solid argument to withhold surgery from a person who fell into the trap of IV drugs. (2) Although physicians should be active members in the conversation of allocating medical resources, they are not solely entitled to make those decisions. (3) Rationing should not take place at patients' bedsides. (4) The solution to this ethical dilemma requires a collective vision of approaching social justice in light of limited medical resources. Therefore, the answers will depend on culture, economics, and many political and social factors and might differ between various countries.

Thought Box (V)

The reductionist approach to studying chronic illnesses is, at the very least, insufficient and, in fact, many times problematic. When we deliver care, we realize that there is a need for the expertise of different specialists in medicine, along with psychiatrists, chaplains, social workers, and many other healthcare professionals. I recently reflected on an interesting phenomenon of branching out fields from cardiology, such as cardio-oncology, cardio-palliative, and critical care cardiology. And indeed, we will continue to have more branches growing from the main trunk. In a sense, those fields expand to reach outsider fields rather than subdividing cardiology into valvular, interventional, electrophysiology, and so on. It appears that is a countermovement on two layers. First, from the healthcare professional standpoint, some physicians prefer a practice with a broader base and would rather keep their knowledge outside a focused specialty. The second layer is more of a model of thinking. What I am referring to is that the phenomenon manifests the need to understand sickness in a broader sense as a disturbed relationship. Fixing an organ will not always fix the disturbed relationship. And keeping in mind this approach is pivotal for offering better care, finding novel research ideas, and getting out of our narrow comfort zones as specialists. Sometimes, we tend to forget that the heart, the kidney, or the liver are part of the whole human being, and that patient care should be offered in light of that. How we will find the balance between the myopic and hyperopic visions of illness is a challenging question, especially while practicing in a healthcare system that is economically based on specialization. However, each physician should strive to remember the need for both approaches as each gives a different kind of knowledge about a patient and, consequently, a broader form of understanding and care.

Notes

1. Hunt SA. Acc/aha 2005 guideline update for the diagnosis and management of chronic heart failure in the adult: A report of the american college of cardiology/american heart association task force on practice guidelines (writing committee to update the 2001 guidelines for the evaluation and management of heart failure) J Am Coll Cardiol. 2005;46:e1–82.
2. Metra Marco, Teerlink John R. Heart failure. The Lancet. 2017;390(10106):1981.
3. Jessup M, Brozena S. Heart failure. N Engl J Med. 2003;348:2007–2018.
4. Metra Marco, Teerlink John R. Heart failure. The Lancet. 2017;390(10106):1983–1984.
5. Ashley K, Cho L. Manual of Cardiovascular Medicine. 3rd ed. Philadelphia: Lippincott Williams & Wilkins; 2009, P120.

6. Cazeau, S., Leclercq, C., Lavergne, T., Walker, S., Varma, C., Linde, C., Garrigue, S., Kappenberger, L., Haywood, G.A., Santini, M., Bailleul, C., Daubert, J.C., and Multi-site Stimulation in Cardiomyopathies (MUSTIC) Study Investigators. Effects of multi-site biventricular pacing in patients with heart failure and intraventricular conduction delay. *N Engl J Med*. 2001; 344: 873–880. Bristow, M.R., Saxon, L.A., Boehmer, J., Krueger, S., Kass, D.A., De Marco, T., Carson, P., DiCarlo, L., DeMets, D., White, B.G., DeVries, D.W., Feldman, A.M., and Comparison of Medical Therapy, Pacing, and Defibrillation in Heart Failure (COMPANION) Investigators. Cardiac-resynchronization therapy with or without an implantable defibrillator in advanced chronic heart failure. *N Engl J Med*. 2004; 350: 2140–2150.

7. Rose EA, Gelijns AC, Moskowitz AJ, et al. Long-term use of a left ventricular assist device for end-stage heart failure. N Engl J Med 2001; 345:1435–43.

8. Benjamin EJ, Blaha MJ, Chiuve SE, Cushman M, Das SR, Deo R, et al. Heart disease and stroke statistics—2017 update: a report from the American Heart Association. Circulation. 2017;135(10):e146–603.

9. Gelfman, L. P., Bakistas, M., Warner Stevenson, L., Kirkpatrick, J. N., & Goldstein, N. E. (2017). The state of the science on integrating palliative care in heart failure. Journal of Palliative Medicine, 20(6), 592–593.

10. Roger VL. Epidemiology of heart failure. Circ Res 2013;113:650.

11. Chen-Scarabelli C, Saravolatz L, Hirsh B, Agrawal P, Scarabelli TM. Dilemmas in end-stage heart failure. J Geriatr Cardiol. 2015; 12:57–65. [PubMed: 25678905].

12. Lloyd-Jones DM, Larson MG, Leip EP, et al. Lifetime risk for developing congestive heart failure: the Framingham Heart Study. Circulation 2002;106:3068–72.

13.. ESC Guidelines. Guidelines for the diagnosis and treatment of chronic heart failure. Eur Soc Cardiol 2005.

14. Setoguchi S, Glynn RJ, Stedman M, et al.: Hospice, opiates, and acute care service use among the elderly before death from heart failure or cancer. Am Heart J 2010;160:139–144.

15. Brännström M, Boman K. Effects of person-centred and integrated chronic heart failure and palliative home care. PREFER: a randomized controlled study. Eur J Heart Fail 2014;16:1142–1151. Rogers JG, Patel CB, Mentz RJ, Granger BB, Steinhauser KE, Fiuzat M, Adams PA, Speck A, Johnson KS, Krishnamoorthy A, Yang H, Anstrom KJ, Dodson GC, Taylor DH Jr, Kirchner JL, Mark DB, O'Connor CM, Tulsky JA. Palliative care in heart failure: the PA.

16. Lum HD, Jones J, Lahoff D, Allen LA, Bekelman DB, Kutner JS, Matlock DD (2015) Unique challenges of hospice for patients with heart failure: a qualitative study of hospice clinicians. Am Heart J 170:524–530.

17. Spiess JL. Hospice in heart failure: why, when, and what then? Heart Fail Rev. 2017;22(5):593–604.

18. Allen LA, Yager JE, Funk MJ et al. (2008) Discordance between patient-predicted and model-predicted life expectancy among ambulatory patients with heart failure. JAMA 299:2533–2542.

19. Sood A, Dobbie K, Wilson Tang WH. Palliative care in heart failure. Curr Treat Options Cardiovasc Med. 2018;20(5):43.

20. Rose EA, Gelijns AC, Moskowitz AJ, et al. Long-term use of a left ventricular assist device for end-stage heart failure. N Engl J Med 2001; 345:1435–43.

21. Cassell, Eric J. "The person in medicine." *International Journal of Integrated Care* 10, no. 5 (2010).

22. Dostoevsky (August 22, 2001), Crime and Punishment, Dover Publications; Reprint edition. P 1–2.

23. Dostoevsky (August 22, 2001), Crime and Punishment, Dover Publications; Reprint edition. P 205–206.

24. Miguel de Cervantes Saavedra (February 5, 2013), Don Quixote, Signet; Abridged edition. P 1–2.

25. Cassell, E. (2004). The nature of suffering and the goals of medicine, second edition, P 164.

26.   Cassell, E. (2004). The nature of suffering and the goals of medicine, second edition P 166–171.
27.   Cassell, E. (2004). The nature of suffering and the goals of medicine, second edition 172–181.
28.   Cassell, E. (2004). The nature of suffering and the goals of medicine, second edition 181–199.
29.   Cassell, E. (2004). The nature of suffering and the goals of medicine, second edition. P 46–48.
30.   Kleinman A. Suffering, healing and the human condition. Encyclopedia of Human Biology. 1988. P4.
31.   Cassell, E. (2004). The nature of suffering and the goals of medicine, second edition. P 49–59.
32.   Kleinman A. Suffering, healing and the human condition. Encyclopedia of Human Biology. 1988.P7.
33.   Sulmasy DP. A biopsychosocial-spiritual model for the care of patients at the end of life. The gerontologist. 2002 Oct 1;42(suppl_3):24.
34.   Sulmasy DP. Is medicine a spiritual practice?. Academic medicine: journal of the Association of American Medical Colleges. 1999 Sep;74(9):1002.
35.   Sulmasy, Daniel P. A Balm for Gilead: Meditations on Spirituality and the Healing Arts [Uncorrected Advance Proofs]. Uncorr Uncorr. advance proofs. ed. Washington, D.C.: Georgetown University Press, 2006. P 22.
36.   Sulmasy DP. A biopsychosocial-spiritual model for the care of patients at the end of life. The gerontologist. 2002 Oct 1;42(suppl_3):25–26.
37.   Lagay F. The legacy of humoral medicine. AMA Journal of Ethics. 2002 Jul 1;4(7).
38.   Earl E. Shelp (1983). The Clinical Encounter, The moral fabric of the patient – Physician relationship. P5.
39.   Sulmasy DP. The rebirth of the clinic: An introduction to spirituality in health care. Georgetown University Press; 2006 May 19. P 125.
40.   Sulmasy DP. The rebirth of the clinic: An introduction to spirituality in health care. Georgetown University Press; 2006 May 19. P 126.
41.   Cassell, E. (2004). The nature of suffering and the goals of medicine, second edition, P 243–244.
42.   Cassell, E. (2004). The nature of suffering and the goals of medicine, second edition, P 236–237.
43.   Kleinman A. Suffering, healing and the human condition. Encyclopedia of Human Biology. 1988. P 29.
44.   Kleinman A. Suffering, healing and the human condition. Encyclopedia of Human Biology P 20.
45.   Sundaram V, Fang JC. Gastrointestinal and liver issues in heart failure. Circulation. 2016;133:1696.
46.   Hillege HL, Nitsch D, Pfeffer MA, Swedberg K, McMurray JJ, Yusuf S, Granger CB, Michelson EL, Ostergren J, Cornel JH, de Zeeuw D. Renal function as a predictor of outcome in a broad spectrum of patients with heart failure. Circulation. 2006 Feb 7;113(5):671–678.
47.   Rangaswami J, Bhalla V, Blair JE, Chang TI, Costa S, Lentine KL, Lerma EV, Mezue K, Molitch M, Mullens W, Ronco C. Cardiorenal syndrome: classification, pathophysiology, diagnosis, and treatment strategies: a scientific statement from the American Heart Association. Circulation. 2019 Apr 16;139(16):e843–844.
48.   Sundaram V, Fang JC. Gastrointestinal and liver issues in heart failure. Circulation. 2016 Apr 26;133(17):1697–1698.
49.   Mozos I. Mechanisms linking red blood cell disorders and cardiovascular diseases. BioMed research international. 2015: 1–2.
50.   Sullivan M, Levy WC, Russo JE, Spertus JA. Depression and health status in patients with advanced heart failure: a prospective study in tertiary care. Journal of cardiac failure. 2004 Oct 1;10(5):390–396.
51.   Müller-Tasch T, Frankenstein L, Holzapfel N, Schellberg D, Löwe B, Nelles M, Zugck C, Katus H, Rauch B, Haass M, Jünger J. Panic disorder in patients with chronic heart failure. Journal of psychosomatic research. 2008 Mar 1;64(3):299–303.

52. Frasure-Smith N. Atrial Fibrillation and Congestive Heart Failure Investigators: Elevated depression symptoms predict long-term cardiovascular mortality in patients with atrial fibrillation and heart failure. Circulation. 2009;120:134–140.

53. Alhurani AS, Dekker RL, Abed MA, Khalil A, Al Zaghal MH, Lee KS, Mudd-Martin G, Biddle MJ, Lennie TA, Moser DK. The association of co-morbid symptoms of depression and anxiety with all-cause mortality and cardiac rehospitalization in patients with heart failure. Psychosomatics. 2015 Jul 1;56(4):371–380.

54. Freedland KE, Carney RM, Rich MW, Steinmeyer BC, Rubin EH. Cognitive behavior therapy for depression and self-care in heart failure patients: a randomized clinical trial. JAMA internal medicine. 2015 Nov 1;175(11):1773–1782.

55. O'Connor CM, Jiang W, Kuchibhatla M, Silva SG, Cuffe MS, Callwood DD, Zakhary B, Stough WG, Arias RM, Rivelli SK, Krishnan R. Safety and efficacy of sertraline for depression in patients with heart failure: results of the SADHART-CHF (Sertraline Against Depression and Heart Disease in Chronic Heart Failure) trial. Journal of the American College of Cardiology. 2010 Aug 24;56(9):692–699.Angermann CE, Gelbrich G, Störk S, Fallgatter A, Deckert J, Faller H, Ertl G, MOOD-HF Investigators. Rationale and design of a randomised, controlled, multicenter trial investigating the effects of selective serotonin re-uptake inhibition on morbidity, mortality and mood in depressed heart failure patients (MOOD-HF). European journal of heart failure. 2007 Dec;9(12):1212–1222.

56. Bury M. Chronic illness as biographical disruption. Sociology of health & illness. 1982 Jul;4(2):169.

57. Siegrist J, Siegrist K, Weber I. Sociological concepts in the etiology of chronic disease: the case of ischemic heart disease. Social Science & Medicine. 1986 Jan 1;22(2):249.

58. Krumholz HM, Butler J, Miller J, Vaccarino V, Williams CS, Mendes de Leon CF, Seeman TE, Kasl SV, Berkman LF. Prognostic importance of emotional support for elderly patients hospitalized with heart failure. Circulation. 1998 Mar 17;97(10):958–964.

59. Chin MH, Goldman L. Correlates of early hospital readmission or death in patients with congestive heart failure. The American journal of cardiology. 1997 Jun 15;79(12):1640–1644.

60. Brännström M, Ekman I, Boman K, Strandberg G. Being a close relative of a person with severe, chronic heart failure in palliative advanced home care–a comfort but also a strain. Scandinavian journal of caring sciences. 2007 Sep;21(3):342–343.

61. Hwang B, Fleischmann KE, Howie-Esquivel J, Stotts NA, Dracup K. Caregiving for patients with heart failure: impact on patients' families. American Journal of Critical Care. 2011 Nov 1;20(6):431–442.

62. Doherty LC, Fitzsimons D, McIlfatrick SJ. Carers' needs in advanced heart failure: A systematic narrative review. European journal of cardiovascular nursing. 2016 Jun;15(4):203–212.

63. Sulmasy DP. The healer's calling: A spirituality for physicians and other health care professionals. Paulist Press; 1997. P 10.

64. Sulmasy DP. The healer's calling: A spirituality for physicians and other health care professionals. Paulist Press; 1997.P 11.

65. Sulmasy DP. The rebirth of the clinic: An introduction to spirituality in health care. Georgetown University Press; 2006 May 19. P 124.

66. Sulmasy DP. The healer's calling: A spirituality for physicians and other health care professionals. Paulist Press; 1997.P 100–101.

67. Sulmasy DP. The healer's calling: A spirituality for physicians and other health care professionals. Paulist Press; 1997.P 98–99.

68. Sulmasy DP. The healer's calling: A spirituality for physicians and other health care professionals. Paulist Press; 1997.P 94–95.

69. Sulmasy DP. The rebirth of the clinic: An introduction to spirituality in health care. Georgetown University Press; 2006 May 19. P 198.

70. Sulmasy DP. The rebirth of the clinic: An introduction to spirituality in health care. Georgetown University Press; 2006 May 19. P 200–201.

71. Sulmasy DP. The rebirth of the clinic: An introduction to spirituality in health care. Georgetown University Press; 2006 May 19. P 210.

72.   Sulmasy DP. The rebirth of the clinic: An introduction to spirituality in health care. Georgetown University Press; 2006 May 19. P 207.

73.   Sulmasy DP. The rebirth of the clinic: An introduction to spirituality in health care. Georgetown University Press; 2006 May 19. P 206.

74.   Sulmasy DP. The rebirth of the clinic: An introduction to spirituality in health care. Georgetown University Press; 2006 May 19. P 205.

75.   Sulmasy DP. A balm for Gilead: Meditations on spirituality and the healing arts. Georgetown University Press; 2006.P 23–24.

76.   Sulmasy DP. The healer's calling: A spirituality for physicians and other health care professionals. Paulist Press; 1997. P104.

77.   Ross, Linda, and Jackie Miles. "Spirituality in heart failure: a review of the literature from 2014 to 2019 to identify spiritual care needs and spiritual interventions." Current Opinion in Supportive and Palliative Care 14, no. 1 (2020): 9–18. Page 9.

78.   Ross, Linda, and Jackie Miles. "Spirituality in heart failure: a review of the literature from 2014 to 2019 to identify spiritual care needs and spiritual interventions." Current Opinion in Supportive and Palliative Care 14, no. 1 (2020): 9–18. Page 16–17.

79.   Murray, Scott A., Marilyn Kendall, Elizabeth Grant, Kirsty Boyd, Stephen Barclay, and Aziz Sheikh. "Patterns of social, psychological, and spiritual decline toward the end of life in lung cancer and heart failure." Journal of pain and symptom management 34, no. 4 (2007): 393–402.

80.   Murray, Scott A., Marilyn Kendall, Kirsty Boyd, Allison Worth, and T. Fred Benton. "Exploring the spiritual needs of people dying of lung cancer or heart failure: a prospective qualitative interview study of patients and their carers." Palliative medicine 18, no. 1 (2004): 39–45.

81.   Clark, Clayton C., and Jennifer Hunter. "Spirituality, spiritual well-being, and spiritual coping in advanced heart failure: Review of the literature." Journal of Holistic Nursing 37, no. 1 (2019): 56–73.

82.   Sulmasy DP. Is medicine a spiritual practice?. Academic medicine: journal of the Association of American Medical Colleges. 1999 Sep;74(9):1003–1004.

83.   Wallace, S. M., B. I. Walton, R. K. Kharbanda, R. Hardy, A. P. Wilson, and R. H. Swanton. "Mortality from infective endocarditis: clinical predictors of outcome." Heart 88, no. 1 (2002): 53.

84.   Murdoch, David R., G. Ralph Corey, Bruno Hoen, José M. Miró, Vance G. Fowler, Arnold S. Bayer, Adolf W. Karchmer et al. "Clinical presentation, etiology, and outcome of infective endocarditis in the twenty-first century: the International Collaboration on Endocarditis–Prospective Cohort Study." Archives of internal medicine 169, no. 5 (2009): 463–473.

85.   Wurcel, Alysse G., Jordan E. Anderson, Kenneth KH Chui, Sally Skinner, Tamsin A. Knox, David R. Snydman, and Thomas J. Stopka. "Increasing infectious endocarditis admissions among young people who inject drugs." In Open forum infectious diseases, vol. 3, no. 3, p. ofw157. Oxford University Press, 2016.

86.   Kim, Joon Bum, Julius I. Ejiofor, Maroun Yammine, Masahiko Ando, Janice M. Camuso, Ilan Youngster, Sandra B. Nelson et al. "Surgical outcomes of infective endocarditis among intravenous drug users." The Journal of thoracic and cardiovascular surgery 152, no. 3 (2016): 832–841.

87.   Wurcel, Alysse G., Jordan E. Anderson, Kenneth KH Chui, Sally Skinner, Tamsin A. Knox, David R. Snydman, and Thomas J. Stopka. "Increasing infectious endocarditis admissions among young people who inject drugs." In Open forum infectious diseases, vol. 3, no. 3, p. ofw157. Oxford University Press, 2016.

88.   Baddour LM, Wilson WR, Bayer AS, et al.; American Heart Association Committee on Rheumatic Fever, Endocarditis, and Kawasaki Disease of the Council on Cardiovascular Disease in the Young, Council on Clinical Cardiology, Council on Cardiovascular Surgery and Anesthesia, and Stroke Council. Infective endocarditis in adults: diagnosis, antimicrobial therapy, and management of complications: a scientific statement for healthcare professionals from the American Heart Association. Circulation 2015; 132:1435–86.

89. Shrestha, Nabin K., Jennifer Jue, Syed T. Hussain, Jason M. Jerry, Gosta B. Pettersson, Venu Menon, Jose L. Navia, Amy S. Nowacki, and Steven M. Gordon. "Injection drug use and outcomes after surgical intervention for infective endocarditis." *The Annals of thoracic surgery* 100, no. 3 (2015): 875–882.
90. Kaiser, Scott P., Spencer J. Melby, Andreas Zierer, Richard B. Schuessler, Marc R. Moon, Nader Moazami, Michael K. Pasque, Charles Huddleston, Ralph J. Damiano Jr, and Jennifer S. Lawton. "Long-term outcomes in valve replacement surgery for infective endocarditis." *The Annals of thoracic surgery* 83, no. 1 (2007): 30–35.
91. Hall, Ryan, Michael Shaughnessy, Griffin Boll, Kenneth Warner, Helen W. Boucher, Raveendhara R. Bannuru, and Alysse G. Wurcel. "Drug Use and Postoperative Mortality Following Valve Surgery for Infective Endocarditis: A Systematic Review and Meta-analysis." *Clinical Infectious Diseases* (2018).
92. Rosenthal, Elana S., Adolf W. Karchmer, Jesse Theisen-Toupal, Roger Araujo Castillo, and Chris F. Rowley. "Suboptimal addiction interventions for patients hospitalized with injection drug use-associated infective endocarditis." *The American journal of medicine* 129, no. 5 (2016): 481–485.
93. Smyth, Bobby P., J. Barry, E. Keenan, and K. Ducray. "Lapse and relapse following inpatient treatment of opiate dependence." *Ir Med J* 103, no. 6 (2010): 176–179.
94. Yeo, Khung-Keong, W. J. Chang, Jeffrey M. Lau, and Siang-Yong Tan. "Valve replacement in endocarditis: setting limits in noncompliant intravenous drug abusers." *Hawaii medical journal* 65, no. 6 (2006): 170.
95. DiMaio, J. Michael, Tomas A. Salerno, Ron Bernstein, Katia Araujo, Marco Ricci, and Robert M. Sade. "Ethical obligation of surgeons to noncompliant patients: can a surgeon refuse to operate on an intravenous drug-abusing patient with recurrent aortic valve prosthesis infection?." *The Annals of thoracic surgery* 88, no. 1 (2009): 3.
96. Strathdee, Steffanie A., and Jamila K. Stockman. "Epidemiology of HIV among injecting and non-injecting drug users: current trends and implications for interventions." *Current HIV/AIDS Reports* 7, no. 2 (2010): 99–106.
97. Lakbala, Parvin, Ghasem Sobhani, Mahboobeh Lakbala, Kavoos Dindarloo Inaloo, and Hamid Mahmoodi. "Sharps injuries in the operating room." *Environmental health and preventive medicine* 19, no. 5 (2014): 348.
98. Lakbala, Parvin, Ghasem Sobhani, Mahboobeh Lakbala, Kavoos Dindarloo Inaloo, and Hamid Mahmoodi. "Sharps injuries in the operating room." *Environmental health and preventive medicine* 19, no. 5 (2014): 291.
99. Baldassarri, Stephen R., Ike Lee, Stephen R. Latham, and Gail D'Onofrio. "Debating Medical Utility, Not Futility: Ethical Dilemmas in Treating Critically Ill People Who Use Injection Drugs." *The Journal of Law, Medicine & Ethics* 46, no. 2 (2018): 247.
100. DiMaio, J. Michael, Tomas A. Salerno, Ron Bernstein, Katia Araujo, Marco Ricci, and Robert M. Sade. "Ethical obligation of surgeons to noncompliant patients: can a surgeon refuse to operate on an intravenous drug-abusing patient with recurrent aortic valve prosthesis infection?." *The Annals of thoracic surgery* 88, no. 1 (2009): 1.
101. Baldassarri, Stephen R., Ike Lee, Stephen R. Latham, and Gail D'Onofrio. "Debating Medical Utility, Not Futility: Ethical Dilemmas in Treating Critically Ill People Who Use Injection Drugs." *The Journal of Law, Medicine & Ethics* 46, no. 2 (2018): 242.
102. Fitzgerald, Faith T. "The tyranny of health." (1994): 196–198.
103. Šarić, Lenko, Ivana Prkić, and Marko Jukić. "Futile treatment—A review." *Journal of bioethical inquiry* 14, no. 3 (2017): 330.
104. Aghabarary, Maryam, and Nahid Dehghan Nayeri. "Medical futility and its challenges: a review study." *Journal of medical ethics and history of medicine* 9 (2016). P 4.
105. Schneiderman, Lawrence J., Nancy S. Jecker, and Albert R. Jonsen. "Medical futility: its meaning and ethical implications." *Annals of internal medicine* 112, no. 12 (1990): 949–954.
106. Schneiderman, Lawrence J., Nancy S. Jecker, and Albert R. Jonsen. "Medical futility: response to critiques." *Annals of Internal Medicine* 125 (1996): 672.

107. Kelly, David F., Gerard Magill, and Henk Ten Have. *Contemporary Catholic health care ethics.* Georgetown University Press, 2013. P 222–223.
108. Kelly, David F., Gerard Magill, and Henk Ten Have. *Contemporary Catholic health care ethics.* Georgetown University Press, 2013. 219–221.
109. Helft, Paul R., Mark Siegler, and John Lantos. "The rise and fall of the futility movement." (2000): 293–296.
110. Schneiderman, Lawrence J., and Nancy S. Jecker. "Should a criminal receive a heart transplant? Medical justice vs. societal justice." *Theoretical Medicine* 17, no. 1 (1996): 34–37.
111. Schneiderman, Lawrence J., and Nancy S. Jecker. "Should a criminal receive a heart transplant? Medical justice vs. societal justice." *Theoretical Medicine* 17, no. 1 (1996): 38–41.

Bibliography

1. Aghabarary, Maryam, and Nahid Dehghan Nayeri. 2016. Medical futility and its challenges: A review study. *Journal of Medical Ethics and History of Medicine* 9: 4.
2. Alhurani, A.S., R.L. Dekker, M.A. Abed, et al. 2015. The association of co-morbid symptoms of depression and anxiety with all-cause mortality and cardiac rehospitalization in patients with heart failure. *Psychosomatics* 56: 371–380 [PubMed: 25556571].
3. Allen, L.A., J.E. Yager, M.J. Funk, et al. 2008. Discordance between patient-predicted and model-predicted life expectancy among ambulatory patients with heart failure. *JAMA* 299: 2533–2542.
4. Ashley, K., and L. Cho. 2009. *Manual of cardiovascular medicine*, 3rd ed. Philadelphia: Lippincott Williams & Wilkins.
5. Baddour, L. M., W. R. Wilson, A. S. Bayer, et al. 2015. American Heart Association Committee on Rheumatic Fever, Endocarditis, and Kawasaki Disease of the Council on Cardiovascular Disease in the Young, Council on Clinical Cardiology, Council on Cardiovascular Surgery and Anesthesia, and Stroke Council. Infective endocarditis in adults: diagnosis, antimicrobial therapy, and management of complications: a scientific statement for healthcare professionals from the American Heart Association. *Circulation* 132: 1435–86.
6. Baldassarri, Stephen R., Ike Lee, Stephen R. Latham, and Gail D'Onofrio. 2018. Debating medical utility, not futility: Ethical dilemmas in treating critically Ill people who use injection drugs. *The Journal of Law, Medicine & Ethics* 46 (2): 247.
7. Baldassarri, Stephen R., Ike Lee, Stephen R. Latham, and Gail D'Onofrio. 2018. Debating medical utility, not futility: Ethical dilemmas in treating critically Ill people who use injection drugs. *The Journal of Law, Medicine & Ethics* 46 (2): 242.
8. Benjamin, E.J., M.J. Blaha, S.E. Chiuve, M. Cushman, S.R. Das, R. Deo, et al. 2017. Heart disease and stroke statistics—2017 update: A report from the American Heart Association. *Circulation* 135 (10): e146-603.
9. Brännström, M., and K. Boman. 2014. Effects of person-centered and integrated chronic heart failure and palliative home care. PREFER: A randomized controlled study. *European Journal of Heart Failure* 16:1142–1151.
10. Bristow, M. R., L. A. Saxon, J. Boehmer, S. Krueger, D. A. Kass, T. De Marco, P. Carson, L. DiCarlo, D. DeMets, B. G. White, D. W. DeVries, A. M. Feldman and Comparison of Medical Therapy, Pacing, and Defibrillation in Heart Failure (COMPANION) Investigators. 2004. Cardiac-resynchronization therapy with or without an implantable defibrillator in advanced chronic heart failure. *New England Journal of Medicine* 350: 2140–2150.
11. Cassell, E. 2004. *The nature of suffering and the goals of medicine*, 2nd edn.
12. Cassell, E.J. 2010. The person in medicine. *International Journal of Integrated Care* 10 (Supplement): 50–52.

13. Cazeau, S., C. Leclercq, T. Lavergne, S. Walker, C. Varma, C. Linde, S. Garrigue, L. Kappenberger, G. A. Haywood, M. Santini, C. Bailleul, J. C. Daubert, and Multisite Stimulation in Cardiomyopathies (MUSTIC) Study Investigators. 2001. Effects of multisite biventricular pacing in patients with heart failure and intraventricular conduction delay. *New England Journal of Medicine* 344: 873–880.
14. Chen-Scarabelli, C., L. Saravolatz, B. Hirsh, P. Agrawal, and T.M. Scarabelli. 2015. Dilemmas in end-stage heart failure. *Journal of Geriatric Cardiology* 12: 57–65.
15. DiMaio, J. Michael, Tomas A. Salerno, Ron Bernstein, Katia Araujo, Marco Ricci, and Robert M. Sade. 2009. Ethical obligation of surgeons to noncompliant patients: Can a surgeon refuse to operate on an intravenous drug-abusing patient with recurrent aortic valve prosthesis infection? *The Annals of Thoracic Surgery* 88(1): 3.
16. Dostoevsky, Crime and Punishment, Dover Publications; Reprint edition (August 22, 2001).
17. Earl E. Shelp, The Clinical Encounter, The moral fabric of the patient—Physician relationship 1983.
18. ESC Guidelines. 2005. Guidelines for the diagnosis and treatment of chronic heart failure. *European Society of Cardiology*.
19. Fitzgerald, Faith T. 1994. *The tyranny of health*: 196–198.
20. Frasure-Smith, N., F. Lesperance, M. Habra, et al. 2009. Elevated depression symptoms predict long-term cardiovascular mortality in patients with atrial fibrillation and heart failure. *Circulation* 120: 134–140 [PubMed: 19564557].
21. Gelfman, L.P., M. Bakistas, L. Warner Stevenson, J.N. Kirkpatrick, and N.E. Goldstein. 2017. The state of the science on integrating palliative care in heart failure. *Journal of Palliative Medicine* 20 (6): 592–593.
22. Hall, Ryan, Michael Shaughnessy, Griffin Boll, Kenneth Warner, Helen W. Boucher, Raveendhara R. Bannuru, and Alysse G. Wurcel. 2018. Drug use and postoperative mortality following valve surgery for infective endocarditis: A systematic review and meta-analysis. *Clinical Infectious Diseases*.
23. Helft, Paul R., Mark Siegler, and John Lantos. 2000. *The rise and fall of the futility movement*: 293–296.
24. Hillege, H.L., D. Nitsch, M.A. Pfeffer, et al. 2006. Renal function as a predictor of outcome in a broad spectrum of patients with heart failure. *Circulation* 113: 671–678.
25. Hunt, S. A. 2005. Acc/aha 2005 guideline update for the diagnosis and management of chronic heart failure in the adult: A report of the American college of cardiology/American heart association task force on practice guidelines (writing committee to update the 2001 guidelines for the evaluation and management of heart failure). *Journal of the American College of Cardiology* 46: e1–82.
26. Jessup, M., and S. Brozena. 2003. Heart failure. *New England Journal of Medicine* 348: 2007–2018.
27. Kaiser, Scott P., Spencer J. Melby, Andreas Zierer, Richard B. Schuessler, Marc R. Moon, Nader Moazami, Michael K. Pasque, Charles Huddleston, Ralph J. Damiano Jr, and Jennifer S. Lawton. 2007. Long-term outcomes in valve replacement surgery for infective endocarditis. *The Annals of Thoracic Surgery* 83(1): 30–35.
28. Kelly, David F., Gerard Magill, and Henk Ten Have. 2013. *Contemporary Catholic health care ethics*. Georgetown University Press, 222–223.
29. Kim, Joon Bum, Julius I. Ejiofor, Maroun Yammine, Masahiko Ando, Janice M. Camuso, Ilan Youngster, Sandra B. Nelson, et al. 2016. Surgical outcomes of infective endocarditis among intravenous drug users. *The Journal of Thoracic and Cardiovascular Surgery* 152(3): 832–841.
30. Kleinman, A. 1988. *The illness narratives: suffering, healing and the human condition*. New York: Basic Books.
31. Lagay, F. 2002. *The legacy of humoral medicine*. Virtual Mentor.
32. Lakbala, Parvin, Ghasem Sobhani, Mahboobeh Lakbala, Kavoos Dindarloo Inaloo, and Hamid Mahmoodi. 2014. Sharps injuries in the operating room. *Environmental Health and Preventive Medicine* 19(5): 348.

33. Lloyd-Jones, D.M., M.G. Larson, E.P. Leip, et al. 2002. Lifetime risk for developing congestive heart failure: The Framingham heart study. *Circulation* 106: 3068–3072.
34. Lum, H.D., J. Jones, D. Lahoff, L.A. Allen, D.B. Bekelman, J.S. Kutner, and D.D. Matlock. 2015. Unique challenges of hospice for patients with heart failure: A qualitative study of hospice clinicians. *American Heart Journal* 170: 524–530.
35. Metra, Marco, and John R. Teerlink. 2017. Heart failure. *The Lancet* 390(10106): 1981.
36. Miguel de Cervantes Saavedra. 2013. Don Quixote, Signet; Abridged edition (February 5, 2013).
37. Mozos, I. 2015. Mechanisms linking red blood cell disorders and cardiovascular diseases. *BioMed Research International* 2015: 682054.
38. Muller-Tasch, T., L. Frankenstein, N. Holzapfel, et al. 2008. Panic disorder in patients with chronic heart failure. *Journal of Psychosomatic Research* 64: 299–303 [PubMed: 18291245].
39. Murdoch, David R., G. Ralph Corey, Bruno Hoen, José M. Miró, Vance G. Fowler, Arnold S. Bayer, Adolf W. Karchmer, et al. 2009. Clinical presentation, etiology, and outcome of infective endocarditis in the 21st century: The international collaboration on endocarditis–prospective cohort study. *Archives of Internal Medicine* 169(5): 463–473.
40. Rangaswami, J., V. Bhalla, J. E. A. Blair, T. I. Chang, S. Costa, K. L. Lentine, et al. 2019. Cardiorenal syndrome: Classification, pathophysiology, diagnosis, and treatment strategies: A scientific statement from the American Heart Association. *Circulation*, 843–844
41. Roger, V.L. 2013. Epidemiology of heart failure. *Circulation Research* 113: 650.
42. Rogers, J. G., C. B. Patel, R. J. Mentz, B. B. Granger, K. E. Steinhauser, M. Fiuzat, P. A. Adams, A. Speck, K. S. Johnson, A. Krishnamoorthy, H. Yang, K. J. Anstrom, G. C. Dodson, D. H. Taylor Jr, J. L. Kirchner, D. B. Mark, C. M. O'Connor, J. A. Tulsky. *Palliative care in heart failure: The PA.*
43. Rose, E.A., A.C. Gelijns, A.J. Moskowitz, et al. 2001. Long-term use of a left ventricular assist device for end-stage heart failure. *New England Journal of Medicine* 345: 1435–1443.
44. Rosenthal, Elana S., Adolf W. Karchmer, Jesse Theisen-Toupal, Roger Araujo Castillo, and Chris F. Rowley. 2016. Suboptimal addiction interventions for patients hospitalized with injection drug use-associated infective endocarditis. *The American Journal of Medicine* 129(5): 481–485.
45. Šarić, Lenko, Ivana Prkić, and Marko Jukić. 2017. Futile treatment—A review. *Journal of Bioethical Inquiry* 14 (3): 330.
46. Schneiderman, Lawrence J., and Nancy S. Jecker. 1996. Should a criminal receive a heart transplant? Medical justice vs. societal justice. *Theoretical Medicine* 17 (1): 34–37.
47. Schneiderman, Lawrence J., Nancy S. Jecker, and Albert R. Jonsen. 1990. Medical futility: Its meaning and ethical implications. *Annals of Internal Medicine* 112 (12): 949–954.
48. Setoguchi, S., R.J. Glynn, M. Stedman, et al. 2010. Hospice, opiates, and acute care service use among the elderly before death from heart failure or cancer. *American Heart Journal* 160: 139–144.
49. Shrestha, Nabin K., Jennifer Jue, Syed T. Hussain, Jason M. Jerry, Gosta B. Pettersson, Venu Menon, Jose L. Navia, Amy S. Nowacki, and Steven M. Gordon. 2015. Injection drug use and outcomes after surgical intervention for infective endocarditis. *The Annals of Thoracic Surgery* 100 (3): 875–882.
50. Smyth, Bobby P., J. Barry, E. Keenan, and K. Ducray. 2010. Lapse and relapse following inpatient treatment of opiate dependence. *The Irish Medical Journal* 103(6): 176–179.
51. Sood, A., K. Dobbie, and W.H. Wilson Tang. 2018. Palliative care in heart failure. *Current Treatment Options in Cardiovascular Medicine* 20 (5): 43.
52. Spiess, J.L. 2017. Hospice in heart failure: Why, when, and what then? *Heart Failure Reviews* 22 (5): 593–604.
53. Strathdee, Steffanie A., and Jamila K. Stockman. 2010. Epidemiology of HIV among injecting and non-injecting drug users: Current trends and implications for interventions. *Current HIV/AIDS Reports* 7 (2): 99–106.
54. Sullivan, M., W.C. Levy, J.E. Russo, and J.A. Spertus. 2004. Depression and health status in patients with advanced heart failure: A prospective study in tertiary care. *Journal of Cardiac Failure* 10: 390–396 [PubMed: 15470649].

55. Sulmasy, D. P. 2002. A biopsychosocial-spiritual model for the care of patients at the end of life. *Gerontologist* 42(Spec No 3): 24–33.
56. Sulmasy, D.P. 1999. Is medicine a spiritual practice? *Academic Medicine* 74: 1002.
57. Sulmasy, D.P. 2010. *The rebirth of the clinic: An introduction to spirituality in health care.* Washington: Georgetown University Press.
58. Sulmasy, Daniel P. 2006. A balm for gilead: Meditations on spirituality and the healing arts [Uncorrected Advance Proofs]. Uncorr Uncorr. advance proofs. Washington, D.C.: Georgetown University Press.
59. Sundaram, V., and J.C. Fang. 2016. Gastrointestinal and liver issues in heart failure. *Circulation* 133: 1696–1703.
60. Wallace, S.M., B.I. Walton, R.K. Kharbanda, R. Hardy, A.P. Wilson, and R.H. Swanton. 2002. Mortality from infective endocarditis: Clinical predictors of outcome. *Heart* 88 (1): 53.
61. Wurcel, Alysse G., Jordan E. Anderson, Kenneth KH Chui, Sally Skinner, Tamsin A. Knox, David R. Snydman, and Thomas J. Stopka. 2016. Increasing infectious endocarditis admissions among young people who inject drugs. In *Open forum infectious diseases*, vol. 3, no. 3, p. ofw157. Oxford University Press.
62. Yeo, Khung-Keong, W. J. Chang, Jeffrey M. Lau, and Siang-Yong Tan. 2006. Valve replacement in endocarditis: Setting limits in noncompliant intravenous drug abusers. *Hawaii Medical Journal* 65(6): 170.

Chapter 6
Deactivating Cardiovascular Implantable Electronic Device at the End of Life, an Ethical Dilemma in Cardiovascular Medicine

> *"When one day, in the mists of time, man felt something beating in his chest, then the extraordinary, the marvelous history of the heart had just been born."*
> Noubar Boyadjian, MD

Questions

- Can pacemakers consider an integrated part of human body?
- Identify the differences between physician assisted suicide, euthanasian and withholding life support machines?
- What are the limitations of the principle of autonomy?

6.1 Historical Review

6.1.1 The Iconography and Symbolism of the Heart

The human heart has always been a special organ in our history. Not only as a body part associated with life and death but also as a special symbolism with earthly and divine links. Throughout different cultures and cavillation, in poetry, narratives, and religious rituals, the human heart represented an emblem of love, friendship, intelligence, and piety.[1] From the heart lovers communicated, by the courage of the heart knights fought, and through the wisdom of their hearts, philosophers taught. The etymology of the word heart shares common roots in the Indo-European family. In Greek, the word *Kardia*, from which the medical term Cardiac comes, is identical with the Latin word *Cor*. It is also similar to the Sanskrit word *Hrid* and the Germanic word *Herz*. The literal meaning of the word *Hrid* is to leap, and the heart was referred to the leaper, the organ that leaps in the chest, an expression that's still commonly used in different languages at present.[2]

The earliest mention of the human heart in literature was found in Sumerian poetry: 'In vain hath my heart's blood been shed', Gilgamesh could feel his heart beating with pride. Similar references can also be found in ancient Chinese and Indian

S. Toro, *Introduction to Clinical Ethics: Perspectives from a Physician Bioethicist*, https://doi.org/10.1007/978-3-031-30804-8_6

literature. Also, in Egyptian history, the human heart possessed a unique place. It was considered the center of the human body and the source of consciousness. During the mummification process, the heart was not removed from the body like other viscera. Egyptians believed that after death, the human heart will be weighed on a scale against an ostrich feather. Based on scale pan movement, the dead person might either go up to Osiris or down to the Nether Regions. In ancient Greece, we find similar notions. Plato believed that men had three different souls. The Thymos (the vigorous part of the soul) was located in the heart, and it was responsible for the courageous acts of soldiers and warriors. The intellectual or the higher soul was present in the brain, placed in the closest body part to the heavens, and was responsible for rationality and wisdom. Lastly, the liver was the location of the lower soul which was linked with emotions and bound to nourishment and desires. Plato's student, Aristotle, was more cardio-centric. The human heart, for him, was by far the most important organ. Its location in the center of the body made it the most suitable place for the human soul. For Aristotle, the heart was the source of heat, honor, intelligence, and the center of body movements and perceptions.[3] The brain function, interestingly, was less important and mainly confined to the cooling process of the body and thermoregulation. After Christianity became the main religion in Europe, the human heart took a remarkable spiritual value. The sacred heart of Jesus Christ became a symbol of love and kindness. From the pierced heart of the savior, the blood was shed for the salvation of humankind.[4]

The iconography of the heart did not only have divine elements but also earthly ones. The symbol of the heart can be found on posters, commercials, advertisements, and famously on play cards. The symbol of a heart pierced with an arrow is a well-known referral to romantic love. Stories, poems, and songs connecting love with the heart are encountered in almost all languages.[5] Yet probably the most significant symbolization of the heart is linked to life and death. The cessation of beating and breathing movements were both considered compelling historical signs of death. Since the sixteenth-century, physicians started to understand better the complex physiology of Cardio-Pulmonary circulation and its role in maintaining body functions. It was noted that any interruption in this circuit would lead to a decrease oxygen supply to the brain, followed by coma and death.[6] This definition was later challenged in the mid-twentieth-century- by the development of life support machines which led to the state of coma depasse and brain death.

6.1.2   History of Cardiac Pacing

The idea of applying an external electric impulse to treat various diseases has long-standing antiquity in medical history. It goes back to the ancient Roman era when Crampfish's electrical charges were used to treat gout and other painful illnesses.[7] In 1791, the Italian physician Luigi Galvani was the first to discover the relationship between electricity and organic tissues. His experiments on Frogs' cardiac and skeletal muscles formed the foundation of the modern science of electrophysiology.[8]

However, it took until the late nineteenth century for a fully integrated theory of cardiac pacing to be formed. This was achieved through the works of the British physician John Alexander MacWilliam.[9] In 1929 a remarkable and exciting success was achieved. During a medical conference in Sydney, Dr. Mark Lidwill announced his premiere success in resuscitating a newborn with cardiac arrest using an external pacemaker. This case, according to the historian Kirk Jeffery, was the first successful pacing of a human being in medical history. (7)

Around the same time, on the other side of the globe, in New York City, another pioneer physician was working on assembling his pacing device. Dr. Albert Hyman built an external pacemaker in 1932. The "Artificial Pacemaker, "as he named it, was a very primordial yet compelling device. It weighed 7.3 kg, had a hand crank generator and a speed controller. It delivered electric impulses through an intercostal needle directly to the right atrium providing basic pacing support. However, his work did not have broad public acceptance at that time. He faced skepticism and litigation that his pacemaker was interfering with God's will. Sadly, his invention was not adopted by any manufacturing company and ended up not being assembled. This event was probably an early sign of many proceeding conflicts that will be brought up by scientific advancements. It resembled one of the emerging challenges that the human conscious had to take in the long journey of accepting modern technology in medical practice.[10]

After World War II, the public mind was more prepared for rapid progress in science and technology. External pacemakers were rebuilt in the early 1950s. The comprehensive raising knowledge about different cardiac arrhythmia led to rapid advancements in cardiac pacing. Many lives were saved, and patients' longevity dramatically improved. Yet external pacemakers of that time had many problems. They were AC-powered, large in size, and uncomfortable to patients. They also limited mobility and caused skin burns at the electrode site. In 1957, a power outage and tragic death of a baby in Minnesota led Earl Bakken—a hospital equipment engineer—to start working on a pacemaker with an alternative power source. He successfully modified the transistor circuit of a metronome to transmit impulses to the heart rather than to a speaker and brought up to the world the first wearable, battery-powered pacemaker. (10)

The next two impactful milestones in cardiac pacing were the emergence of transvenous pacing techniques and the development of implantable devices. In 1960 a Swedish team led by Dr. Ake Senning implanted the first cardiac pacemaker. Mr. Arne Larrson, who was the first patient to receive this device, ultimately outlived his physician and had more than 20 pacemaker replacements until his death in 2002.[11] Since implanting the first pacemaker in 1960s, significant improvements appeared in pacing technology. Epic advancements were made to the battery types and pacing modalities, including rate-responsive, dual-chamber devices, and intracardiac defibrillators (ICDs).[12] In the early 1990s, Cardiac resynchronization therapy was first described, and its applications later significantly impacted the lives of many patients with heart failure and different arrhythmias.[13] In 2016 leadless pacemakers were invented and currently work is being conducted on manufacturing battery-less cardiac devices.[14] However modern scientific advancements—with pacemaker therapy being one of them—brought up significant challenges to the field of medicine. Not only causing serious morbidity secondary to infections, lead extractions, and posttraumatic stress after ICD shocks, pacing devices also complicated patient's management at the end of life and raised many ethical concerns about deactivating those devices in terminally ill patients.

6.2 General Facts

Since 1960, the implantation of cardiac pacemakers had saved the lives of many patients with fetal arrhythmias.[15] Over the following decades, the applications of cardiac pacing advanced to include treating patients with heart failure, different cardiomyopathies, and lethal tachyarrhythmias.[16] In 2010 it was estimated that more than 2 million patients in the United States have pacemakers, a number that is expected to be much higher at present as the indications for placing therapy widely expanded.[17] However, the implantation of pacemakers led to unexpected and unintended challenges. It increased the number of elderly populations with chronic debilitating illnesses. Significant Anxiety was reported after pacemakers' implantation, and patients had a difficult time coping with the new devices in their bodies. Cardiovascular implantable electronic devices, along with other medical technologies, also indirectly increased the prevalence of cancers, as individuals lived longer to develop such pathologies.[18] Despite their success in treating variant arrhythmias, studies demonstrated an annual mortality rate of 5–20% for recipients.[19] Many patients reached the end of life having implanted devices in their bodies. Henceforth, physicians found themselves facing a variety of ethical and practical challenges on how to deal with these devices in terminally ill patients.

Studies have shown that clinicians are less comfortable with forging pacemaker treatment compared to other life support therapies. A survey of electrophysiologists revealed that despite extensive personal experience and frequently performing deactivations, many remained uncomfortable with withdrawing devices and preferred to involve psychiatrists and ethicists.[20] In reviewing the literature, it is clear that the views on deactivating device therapy vary between pacemakers and ICDs. The opinions also vary whether patients are totally dependent on pacemaker treatments or not. In a survey done in Mayo Clinic in 2010 on the perspective of withdrawing pacemakers and ICD therapy at the end of life, 31% of the physicians considered deactivating pacemakers in a pacemaker-dependent patient to be akin to physician-assisted suicide. The same survey revealed that physicians were more likely to perceive deactivating ICDs as legal compared to pacemakers (85% versus 41%; P < 0.001).[21] Many other studies showed similar results with respondents being more comfortable with deactivating ICDs rather than pacemakers.[22] These statistics reveal that many physicians experience heavy anxiety while taking such decisions, especially in PPM-dependent patients.

The same concerns can be applied to the other side of the equation, namely the patient. It can also be applied to the surrogate decision-makers who are in many times directly responsible for making the decision.[23] Pacemaker function is quite complicated, and many patients do not fully understand the indications for having their devices.[24] Previous works showed that patients suffer from heavy emotional and psychological burdens after ICD implantation. Fear of having shocks, significant anxiety about device malfunction, and body image distortion all have been reported.[25] Studies also have shown that patients do not discuss how to manage their devices at the end of life. In a survey of 278 patients done in an academic tertiary care center, 86% of the respondents had not considered what to do with their ICDs if they had a terminal illness.[26] Same survey showed that only three patients had their wishes about the device included in the advance directives. This latter point has been confirmed in many other studies.[27] Lastly, similar to physicians, patients' views on withdrawing devices therapy at the end of life varied between pacemakers and ICDs. While many agreed with turning off ICDs at the end of life to avoid shocks, others considered the same action to be akin to Euthanasia in case of pacemaker deactivation. (21)

6.3 What is Special About CIEDs?

As mentioned before, previous work showed that physicians are more reluctant to withdrawal pacemaker treatment compared to ventilator and dialysis machines (19). Some scholars tried to identify the driving causes and to point out the unique characteristics which made these devices different from other forms of life-sustaining treatments.

6.3.1 Ontological Analysis

What is unique about CIEDs from an ontological standpoint of view? In general, the special properties of pacemakers can be broadly divided into two main categories. The first is *Anatomical,* more specifically the presence of the device inside the organism and if that makes a moral difference. The second is *Physiological* related to the type of the delivered therapy and if that makes a moral difference. Both categories raise a reasoning question of whether to consider the pacemaker an integral part of the human body? Should the device be recognized as a "body organ" similar to the heart and lung or should it be considered an external live sustaining machine, like ventilators and ECMOs? Although it is a challenging task to draw a clear line between the two sets of properties, namely the *Anatomical* and *Physiological*, it is worth trying to discuss their effects separately.[28]

Regarding the *Anatomical*, it is less problematic and controversial if the discussion was only confined to the "location" itself. In his paper on the deactivation of ICDs, Goldstein pointed out to two main effects that the internalized device had on physicians.[29] First, they were reluctant to discuss device therapy because these small devices had less impact on the quality of patients' lives compared for example to large ventilator machines; a patient connected to a large ventilator by a plastic tube in the mouth would be more burdened compared to another with a small swelling under the skin caused by a pacemaker. Second the fact that the device is small and implanted reminded physicians less to discuss its management as the physical object was not always visible. However, those two points probably are not the main concerns that physicians have. Clearly, no moral argument can be directly derived if the device is implanted inside the body from the anatomical sense alone. To influence the ethical argument, a functional component of the device should be perceived in the biological and psychological sense as a part of human body, and so to speak, it supports the organism functions as an integrated unified organ.

Does pacemaker deliver a treatment model that is similar to a transplanted heart? To answer this question, the discussion has to move into the *Physiological* properties of pacemakers. In general, all treatment interventions tend to restore body functions under two main broad types of therapies. They either provide a *Regulative* therapy or a *Constitutive* therapy.[30] *Regulative* therapies work to get the body back towards normal hemostasis and equilibrium. Antipyretics, antibiotics and antiarrhythmic drugs are all common examples of *Regulative* interventions. On the other hand, *Constitutive* therapies tend to either *substitute* or *replace* a lost body function. Hemodialysis machines and insulin pumps are examples of *Substitutive* therapies, whereas transplanted kidneys and hearts are examples of *Replacement* therapies. By broadly dividing therapeutic interventions under these two categories, it seems that discontinuing constitutive therapies—at the end of life—would raise more moral questions compared to regulative ones. (28)

In his landmark paper on this issue, Salmusy suggested the following criteria to consider a therapeutic intervention a *Replacement* therapy (1) It has to restore the normal physiological function in a way similar to the natural one. (2) It has to respond to internal and external stimulation. (3) It has to grow and self-repair. (4) Does not need an external energy source to function. (5) It has to be in-depended on external control (6) and finally immunologically compatible. Based on these principles, someone can identify the difference between a pacemaker and a transplanted heart or between an insulin pump and transplanted islet cells. Although pacemaker restores a part of lost physiology which is the intrinsic cardiac conduction system, its function differs from the native cardiac electric circuit. It is not integrated into the human body; it neither grows nor repairs itself. The location inside the human body does not change that fact. Peritoneal dialysis occurs inside the body, and although it restores a lost physiological function, it is clearly not equivalent to a transplanted kidney.[31]

With the current advancements in medical technology, it will be more challenging to separate between what is really "Bio" and "Techno" in the literal sense. Perhaps that is why pacemakers, artificial hearts, LVADs, and other new biotechnological devices have created such a problem (31). They challenged our understanding of the boundaries of self. [32] They provided similar functions of the malfunctioning organs, yet they are very different in the made-up structure and microanatomy. Neal Kay and Gregory Brittany suggested considering the pacemaker a replacement therapy rather than substitutive. They argued that a functioning pacemaker has many of the characteristics of replacement therapy. It restores the conduction system of the heart, it is located inside the body, and it is immunologically compatible. Pacemaker responds, senses, and adjusts the rate and impulse amplitude independently of an external energy source.[33] Although not many physicians and scholars agree with their analysis, the points they have raised are understandable in the light of the extremely dynamic technology. Some scholars suggest other theories like Bio-fixture analysis and the On–Off bottom concept to answer this ontological question.[34] All these approaches have their strengths and defects.

6.3.2 *Phenomenological Analysis*

The way a patient perceives the device might also have something to say regarding this issue. One of the patients in Goldstein paper described how she talks to her pacemaker. This expression delineates a significant notion. Despite its location inside her body, the patient did not consider the pacemaker part of herself (29). Similarly, this concept was adopted in the argument of Felicitas Kraemer on how LVAD challenges the concept of Euthanasia.[35] She suggested discussing LVADs not only as an internal or external device, in the ontological, but rather to address LVADs on a *Phenomenological* level. For her, we should ask how patients themselves perceive these devices? What are their life experiences with them? And how would they feel about turning them off?[36] She refers to the famous example of Merleau-Ponty of "blind man's cane" and how it was perceived as part of the body for that individual even though it was external and artificial.[37] Kraemer challenged justifying turning off LVADs based only on the Ontological view and suggested adding a phenomenological consult to analysis. However, the author also acknowledged that her phenomenological approach has its own problems. First, it is a descriptive, non-normative approach, and it cannot solely form an ethical guideline. Second, the perception of the device can be recognized differently by physicians and surrogate decision-makers, which might complicate things even more. Lately, it is casuistic, might or might not be generalizable to all cases, as different patients would also perceive it differently. (36)

6.4 Practical Consideration Regarding Deactivating CIEDs

Many clinicians considered deactivating a pacemaker an act akin to physician-assisted suicide. Some went even further to regard the action as active Euthanasia (21). Before diving into the details of the relation between ontological analysis of pacemakers and their effects on approaching "end of life" decisions, it is essential to clarify some common definitions and concepts. *Euthanasia* etymologically comes from the Greek words *Eu* which means "well," and *Thanatos* which means "death." It is defined in the Oxford dictionary as the painless killing of a patient suffering from an incurable disease. In Euthanasia, the physician intends to cause a patient's death by administering a lethal medication. On the other hand, in *Physician-assisted suicide (PAS)* the physician helps a competent patient to end his/her life by providing a lethal method or medication. The patient then uses the method provided to end his/her life. In both PAS and Euthanasia, death occurs directly as a result of the intervention. *Allowing to die*, on the other hand, or as sometimes called passive Euthanasia, is an act in which a physician either removes an intervention or refrains from performing an action that hinders a preexisting fetal condition.[38] The physician, in that case, might or might not intend the patient's death. Ergo the act itself might be morally right or wrong.[39]

Before carrying on in the discussion, it is essential to re-mention an interesting point. Physicians in many studies considered deactivating pacemakers at the end of life to be ethically different from deactivating other forms of life-sustaining interventions like ventilators and artificial nutrition (17). The reasons behind such a difference pacemakers' perception are not very clear. The following analysis will not focus on answering this question. The argument in this section will specifically shed light on the ethical challenges that will follow both *internality* and *externality* approaches. As for now, even if most physicians continue to agree that pacemakers cannot be considered a unified, integrated body organ, the advancements in medical technology might challenge us with new devices that closely resemble our bodies anatomically and physiologically. A bio-technological, self-repairing, energy-producing pacemaker might not be science fiction in the near future. What has been considered a few decades ago fiction is currently being lived as a reality.[40]

6.4.1   Internality Approach to Pacemakers

The internality approach to pacemakers is not confined only to the location of the device, whether it is inside or outside the body, but also whether the pacemaker is functionally considered an integral part of the organism, similar to a transplanted kidney or a transplant heart. Although this proposal might seem appealing to many, adopting this approach raises many ethical concerns. First, if the pacemaker were thought to form an integral unity with the heart, deactivating it at the end of life would be akin to actively killing the patient, committing a form of "active Euthanasia." The act of turning off the pacemaker would technically be equivalent to injecting a patient with potassium chloride and paralyzing his/her heart. The physician, in that case, is no longer dealing with a device or a life-sustaining treatment; he is directly causing a lethal insult to a native body organ. According to Sulmasy, the definition of "Killing "includes committing an act that creates a new lethal pathophysiological state with the intention of causing a person's death. Whether or not the physician is intentionally trying to kill the patient is not the main point of discussion here. The main concern is whether deactivating a pacemaker, in that case, would simply resemble withdrawing artificial nutrition or hemodialysis. Pacemaker based on internality theory ceases to be an artificial device, ergo, deactivating it would be considered "active Euthanasia." Although some would argue that no lethal medication is being introduced in this case to cause patient death, the point here is related directly to the device itself. It is similar to injecting a drug to kill transplanted islet cells or stopping immunosuppressants to induce kidney rejection. The cause of death based on the internality approach would not be the underlying disease but rather a newly introduced pathological state. The second problem with the internality approach has to do with a patient's autonomy. Considering CIEDs to be an integrated part of the natural heart limits the patient's right to refuse treatment on the ground of autonomy. Even if the patient considered hospice in the last days of life, he/she could not ask the physician—for example—to explant a transplanted heart to facilitate death. Patient's autonomy has limits in that case. No physician would agree with the patient's request to explant a transplanted kidney because he/she just no longer wants the treatment. On the other side, it would also be ethically problematic for the physician to consider withdrawing pacemaker therapy on the ground of futility for the same reason. The application of the whole methodology and ethical principles at the end of life would be very challenging.

6.4.2 Externality Approach to Pacemakers

The externality approach to pacemakers is less problematic and more appealing. As mentioned before, most physicians continue to consider CIEDs a form of life-sustaining treatment. The main problem in taking this approach lies in the differences between pacemakers and other forms of life-support devices. How much do CIEDs resemble ventilators, artificial nutrition, and dialysis machines? The analysis here will follow Kreamer's critique of the externality approach to LVADs. Kreamer listed some of the unique characteristics of LVADs that made them different from other forms of life-sustaining treatments. And Pacemakers share many of those features of LVADs. First, they are implanted in the body, probably even more than LVADs in the literal sense of what implanted means. Second, on the contrary to ventilators, pacemakers are owned by the patient, not the hospital. Third, while ventilators are usually situated in hospitals and other forms of health care facilities, pacemakers move with the patient. Patients live, work, drive and eat with these devices outside the hospital.[41] This functionality component is probably one the most persuasive arguments against the externality approach. Depending on the extent of their disability, patients continue to live a normal or a near-normal life with pacemakers and LVADs. Simon elaborated on this point with a hypothetical example of a backpack ventilator.[42] A small movable machine that provides sufficient support to the patient to function independently outside the hospital. In such a case, he argued, it might be necessary to reassess the ethical approach to extubating patients and withdrawing ventilator support.

6.5   Between Autonomy and Futility

The last HRS consensus statement on the management of CIEDs at the end of life emphasized the importance of respecting Autonomy: patients with intact decision-making capacity have a moral and legal right to refuse any medical treatment/intervention, even if such decisions lead to their death.[43] The AMA statement on end-of-life also affirmed this concept: the duty of physicians to honor patients' wishes about withholding/withdrawing life sustaining treatments.[44] Failure to respect those wishes is a battery in patient-doctor relationship, regardless of the physician's intentions. At the same time, the HRS consensus stated that a clinician's beliefs and personal views should also be respected. Physicians should not be compelled and forced to perform or refrain from performing an intervention when these actions conflict with their personal views. In such cases, they suggest, the clinician must not abandon his/her patient but rather engage a colleague who is willing to proceed with the requested treatments (43). The last recommendation highlights a conflict between two fundamental ethical principles, namely Autonomy and Non-maleficence. Conflicts such as these between a patient's wishes and a physician's obligations are not uncommon. Aging, cultural pluralism, financial conditions, and technological advancements all complicated and challenged the traditional model of the relationship between patients and physicians. An isolated simple generalization of an ethical principle is neither practical nor applicable. Hierarchal prioritization is also problematic as none of these principles can take precedence at all times. Instead, they have to be weight against each other in the context of a clinical case.[45] Despite all of that, the pivotal moral concepts of Autonomy, Beneficence, Non-maleficence, and Justice continue to be cornerstones of biomedical ethics and ethicists continue to rely on them in addressing moral dilemmas in clinical practice.

6.5.1   Autonomy

The word autonomy is derived from two words in Greek, *auto*, which means "self, "and *nomos* which means "rule."[46] Initially, it was confined to the political world, referring to the independence and self-rule of a state. The philosophical roots of autonomy extend back to the enlightenment period in England and France during the emergence of the concepts of individual rights and freedom of choice.[47] Based on those notions, the Individual has a right to act freely according to his/her wishes and believes as the government independently controls and rules its own territory (46). In the United States, the concept of autonomy has carried a significant weight since the foundation of the country. The current emphasis on respecting patients' choices is part of that shared public individualistic priority (45). However, the strong influence of autonomy in medical practice is relatively new, and it goes back only to several decades ago. Historically, since Hippocrates's time, paternalism in medicine has been the dominant model for two thousand years. Beneficence and non-maleficence were the two central principles determining the relationship between physicians and patients. Physicians had absolute authority and responsibility in their work. Patients were rarely if at all, engaged in making decisions. Therefore, it was left to the doctor to determine the best plan and treatment for the patient (47). The transition from the authoritative paternalistic model of relationship to an individualistic autonomic form was the result of two main forces: *First,* the spread of democratic movements, which held an intrinsic mistrust to all forms of authority. The media started to spread public fears and highlighted how physicians abused their power, showing, for example, that they cared only about money they did not hesitate to make their patients stay long hours in waiting rooms before seeing them. As a result, the prestigious, prerogative image of the physicians was seriously challenged and questioned. *Second,* the general improvement in public education decreased the knowledge gap between physicians and patients. People started to read and learn more about different diseases, treatments, and medical interventions. As a result, they were more prepared to discuss and get involved in making decisions about their illnesses and treatments.[48] Nevertheless, the classical-iconic figure of the physician continues to be strongly present in the minds of many patients in the current time and probably it is still the dominant image of clinicians and surgeons in the collective conscious of people internationally and worldwide.

Despite all these facts, respecting patients' autonomy remains a complex concept. All theories of autonomy propose two fundamental conditions on the patient: liberty and capacity. The first is the patient's ability to act independently of any control or influence. The second is his/her capacity to choose and decide between different options.[49] Many conditions influence these two components. The fact of illness, for example, significantly limits patients' liberty. When people become sick, they become vulnerable. Sickness subjects patients to anxiety, fear, distress, and pain. Those factors put a significant burden on patients and complicate the decision-making process.[50] That is why many patients ask their physicians to make medical decisions for them, a point emphasized by Simon in this paper on LVAD deactivation (42). The patient in the presented case asked his physicians to turn off the LVAD when he could disconnect the power supply of the device himself.

The second limitation to autonomy is the decision-making capacity of the patient. At certain times this capacity becomes impaired by the state of illness. For example, in patients who are suffering from depression or severe dementia. The absence of surrogate decision makers or clear advance directives complicates the cases even more. In such clinical scenarios, the physician has a moral obligation to act in the best interest of the patient until further knowledge and clarifications of the patient's wishes are available.[51] Finally, the last problem in autonomy has mainly a social dimension. The multicultural society in the United States places the importance of patient's autonomy at different levels in practical clinical ethics.[52] A study in UCLA on the ethnicity and attitudes toward a patient's autonomy showed a significant difference in end-of-life decisions making between variant cultures. While most African and European Americans believed that elderly patients should be told about their terminal disease and get involved in making decisions about their care, Korean and Mexican Americans were less likely to believe that patients should be informed about their prognosis; and the family rather than the patient should make decisions about life support treatment (60 and 65% versus 28 and 41%).[53] Lastly, in clinical practice, patient's moral right to refuse treatments are easier to be honored than the right of requesting treatments, as the latter not only involves the patient but mandate the physician's participation. Patients, for example, can refuse a procedure, treatment or surgery. They can request not to spend the rest of their life attached to a ventilator machine, but they can't demand treatments or interventions at the same authoritative level. Physicians have a moral obligation not to provide a treatment or intervention when it is not clinically indicated.[54]

6.5.2 Futility

The concept of Futility (Pointless treatment) was introduced in the late 1980s for the sake of discontinuing life-supportive treatments at the end of life. Historically, it goes back also to the Hippocratic Oath "*avoid over treating a patient* ". In Webster's dictionary, Futility is defined as "serving no useful purpose; completely ineffective or producing no valuable effect."[55] From this definition, someone might ask three different questions about a futile treatment: (1) What is the goal of treatment? (2) What is the likelihood that the treatment would be successful? And if it did succeed, (3) are the results achieved valuable? The answers to each of these questions vary depending on the clinical case. The goals of treatment, for example, can extend from complete recovery in a young patient with ARDS, to preparing an elderly patient with multi-organ failure for a peaceful death. Learning about patients' goals of care are essential in the decision-making process.[56] Patients, and families sometimes have unrealistic expectations, and they have to be reminded that medicine has tremendous power but not an ultimate one.[57] The answer to the second question has a probabilistic aspect. What is the percentage of success of a given surgery, or what is the rate of response to a specific chemotherapy? Here the concept of weighing burdens and benefits has an important role. Most of the medical interventions and treatments have side effects. Some of them are very serious and life-threatening, while others are more benign. In clinical practice, physicians learn to deal with uncertainties, to recognize disease patterns in different patient's presentations but never absolutely relay on them diagnosis and making treatment decisions. Medicine is hardly quantitative, and even in the presence of large clinical trials supporting a treatment, no absolute certainty can be guaranteed. In clinical trials, a P value of 0.05 indicates that the observation shown has one in twenty chances of being insignificant, even if the statistics indicated that the medication was effective.[58]

The third element in assessing the futility of medical treatments is a qualitative one. Basically, answering the following question: was the patient able to appreciate the achieved benefit? Modern medicine has tremendous power in altering an individual's physiology and replacing the function of many organs. However, many of these minor successes ultimately don't significantly change the disease course or have qualitative benefits on a patient's life. For example, intermittently stabilizing the hemodynamics of a patient with PVS (persisted vegetative state) or lowering the blood pressure by 5 mm/Hg in an outpatient setting. What can be achieved in the literal sense is not always valuable (58). Lastly, it is also worth mentioning that the qualitative assessment of the goals of treatment is essentially subjective. Its importance might differ between patients, physicians, and sometimes families. Prolonging a patient's survival with devastating leukemia for two months using toxic chemotherapy might be futile in the eye of the treating physician but significantly important for a patient waiting for the birth of her granddaughter in four weeks (56). Weighting the risks and benefits of a medical treatment/intervention and understanding the overall circumstances surrounding the clinical case are both essential in evaluating and addressing ethical challenges.

6.6 Case Discussion

Mr. G is an 82-year-old gentleman with a significant medical history of COPD, Type II diabetes, ischemic cardiomyopathy with decreased ejection fraction (EF 15%). Mr. G has been admitted to the hospital many times in the last six months for pneumonia and worsening respiratory failure. He received multiple courses of antibiotics and IV diuresis. In the last three months, Mr. G lost significant weight, and his ability to function at home drastically decreased. Family and caring physicians noticed that Mr. G's health was deteriorating fast, and his quality of life had significantly declined. Unfortunately, to complicate his health condition, work up during the last admission revealed the presence of underlying lung adenocarcinoma. The patient wasn't thought to be a candidate for chemotherapy. The staff team recommended proceeding with paliative route. The patient and family agreed with the plan, and the decision was made to discharge Mr. G to home-hospice care. Mr. G had a CRT-D placed five years before for primary prevention and heart failure treatment. Before discharge, Mr. G asked his primary physician in the hospital if he was willing to turn off his pacemaker. He disclosed that he neither wants to be back in the hospital for any follow-ups nor to receive any painful shocks from the ICD. He just wanted to die in peace at home between the arms of his loved ones.

Is deactivating the ICD on Mr. G's request a morally licit act? Despite the differences between CIEDs and other forms of life-sustaining treatments, most physicians continue to agree on considering CIED a form of a treatment device. The internality approach to pacemakers is significantly problematic. As mentioned before, treating pacemakers as equivalents to transplanted hearts challenge the whole approach to device deactivation. In that case, the patient cannot request deactivating CIED on the ground of autonomy, as this would be akin to active euthanasia or physician-assisted suicide (Imagine a patient with terminal cancer requesting removal of his kidneys or heart to hasten death). Also, the physician can't withdraw the treatment in that case on the ground of futility for the same reason. The functionality analysis is attractive but also problematic. Despite the fact the patient function with the device outside the hospital, that factor alone doesn't make the device an integrated body organ. Patients function with nicotine patches and baclofen pumps, but we can't consider these therapies/devices integrated body organs based on the functionality aspect alone.

The ICD function of the CIED per se without the bradycardia treatment can be considered based on Sulmasy's approach a *Regulative* therapy. It delivers antitachycardia pacing or defibrillates the heart to restore normal body rhythm and hemostasis (31). Pacemakers, on the other side, are considered *Constitutive—Substitutive* therapy. They restore part of lost normal physiological function, but they don't meet the criteria to be considered a *Replacement* therapy. Pacemakers don't grow or self-repair. They also depend on an external energy source and physician expertise to function appropriately. Moreover, ICDs are not immunologically compatible, white blood cells don't infiltrate into the device to fight infections, and the body forms a fibrous reactive sheath around pacemakers treating them as foreign objects. Based on the previous discussion CIEDs can be considered with relative confidence, life-sustaining therapeutic interventions. However, we should keep in mind that in the future, we might be challenged by newer versions of devices where we have to reconsider our approach to deactivating pacemakers.

From an ethical standpoint, the patient has a moral right to request withdrawal of pacemaker and ICD treatment on the ground of autonomy. There is no moral difference between withdrawing or withholding life support treatments. No ethical difference is present between removing an implanted device or refusing to have the device implanted. As long as the patient was deemed to have adequate capacity to make decisions, his requests should be honored.

Removing the ICD function of the device is less problematic for many physicians. In our case, for Mr. G, receiving a shock from the device would be a quite fearful and unpleasant thing. Especially that he wanted to go home to die in peace. Even from a purely clinical standpoint, the benefits of treating V-fib would probably be temporary in light of the known poor prognosis. So, most of the physicians, if not all of them, would agree to withdraw the ICD function of the device. ICD therapy, in this case, can be considered an extraordinary therapy. But what about the CRT-P, the pacing function? The benefit-burden balancing here is not as straightforward. It is largely dependent on the patient's overall expected longevity and clinical condition. Deactivating CRT-P might result in worsening heart failure symptoms, increasing fatigue, dizziness, and shortness of breath. If the patient was PPM-dependent, it might result in instantaneous death. A common misconception among physicians is that pacemakers will prolong the survival of a dying patient. This is not quite accurate. Terminally ill patients usually become acidotic, and the myocardium doesn't respond to the pacing charges.[59] In our case, for example, Mr. G would most likely die from an infection or respiratory failure,[60] in both cases he will end up in respiratory or lactic acidosis, or even both, and his heart will not probably respond to the device pacing charges.

Furthermore, pacemakers are small devices that have no significant physical burden on the patient. The main concern for Mr. G was the follow-ups; he did not want to go back to see a physician to have his device checked. His request for sure should be honored, but this request does not necessitate deactivating the device. In case Mr. G insisted on his device being deactivated, his request for withdrawing the device should be respected, but the physician has to inform him about the burdens that he might experience if the pacing part of the device was stopped. He might struggle with worsening heart failure symptoms. Informing him about the benefits and burdens should be conducted before proceeding with the patient's request. In this presented case, Mr. G was not completely dependent on PPM therapy. He had an underlying rhythm, and his CIED was implanted as part of heart failure management. Requesting CIED deactivation in PPM-dependent patients can be seriously challenging to many physicians. As mentioned before, this action can result in rapid death. From an ethical standpoint, justifying de-activating PPM function will again depend on weighing the burdens and benefits, and this process will vary from case to case. For example, deactivating PPM in a pacemaker-dependent patient who is brain-dead would be different from the same action in a patient with terminal cancer who is expected to live less than six months. What is morally extraordinary in the first might be ordinary in the second.

In the end, it is worth mentioning an important point that might explain the reluctance of many physicians to withdraw pacemaker therapy compared to other life-sustaining treatments. The fact is primarily emotional. The beating heart has always been considered in history a symbol of life. Taking a decision to remove a device that supports that symbol is consciously and subconsciously very burdensome on the physician. Especially when the patient is still awake and communicating but requesting palliative care. The lack of standard definitions of end of life in palliative care literature adds more complexity to this issue. The burdens of deactivating a pacemaker in someone with a life expectancy of weeks might be different from deactivating it in a patient who is expected to live only a few hours. Both are considered terminally ill patients. The approach to our case, for example, might be very different if Mr. G was in PVS and his family requested comfort care. Generalizing the justification of pacemaker withdrawal at the end of life is not practical. Specific factors related to the patient's overall condition and physician's perspective—in each case—should be taken into consideration before finalizing the moral decision about deactivating CIEDs.

6.7 Summary

Deactivating pacemakers at the end of life can be profoundly challenging for many physicians. Literature showed that physicians rarely discuss pacemaker's deactivation with their patients. Device deactivation is also seldomly reported in patients' advance directives. Moreover, some physicians still consider deactivating pacemakers an act akin to euthanasia and physician-assisted suicide. Based on current scientific progress and in light of the available evidence, pacemakers can still be considered a form of life-sustaining therapy, similar to ventilators, LVADs, and dialysis machines. How we might change our approach in the near future is yet to be determined, and it is likely that the conversation will be kept open. Finally, CIEDs deactivation does not follow a straightforward algorithm in all cases. Circumstances around each case must be considered, and patients should be acknowledged to the benefits and burdens following device withdrawal.

Thought Box (VI)

It is essential to remember that the discussion around killing and allowing to die should happen in the context of patient care. This is quite a fundamental point. Bedside care is different from, for example, a war zone situation when it comes to moral reasoning because there is a code of conduct a physician must obey in light of his role. What do I mean by that? Imagine that you are fighting with your friends on a battlefield, a mortar shell falls, three of your friends die immediately, and one is left alive in a severely deformed state, screaming in unbearable pain and bleeding from many wounds. You can't save him, and you are seeing him suffering. Someone could argue that ending his suffering with a bullet might not be the wrong thing. This example is not similar, in context, to injecting potassium and euthanizing a patient suffering from major depression and lack of social support in a small quiet town in North America. But why? First, there is plentitude of ways to help this person rather than ending his life. Even in cases of severe pain, we have many medications that we can use to alleviate physical and psychological pain. Second, there is an established code of conduct for us as physicians to try to save the lives of our patients when we can and not kill them. Imagine what could happen if physicians were allowed to euthanize patients. Imagine the erosion of public trust. What sort of conversation could a clinician have with a family in the ICU when the mom and dad know he is allowed morally and legally to end their daughter's life? If euthanasia is lawfully permitted, it should be practiced by euthanasia specialists and not doctors.

Now back to our main discussion regarding pacemaker deactivation. There are two practical points to be made here. First, the differentiation between killing and allowing to die holds in this situation. By deactivating the device, the physician is not killing but allowing to die. That doesn't mean his action is morally permissible unless the clinical context suggests so. For example, deactivating the pacemaker to end a patient's life based on his request due to severe depression is not morally sound even that the action in such scenario is allowing to die and not killing. However, turning off the pacing in a patient with end-stage brain metastasis, sepsis, and respiratory distress is very different from an ethical standpoint. Second, many of these arguments are quite theoretical. In practice, there is almost always no need to deactivate a pacemaker's pacing component at the end of life. Especially when a patient is under sedation. Most of these patients will be quite acidotic. There will be a lack of capture, and the spikes on the monitor will have no hemodynamic effects. In such scenarios, stopping serial vitals and unnecessary medications, turning off the monitor, and focusing on patient comfort are preferable actions, and many institutions follow those practices during end-of-life sedation.

Notes

1. Boyadjian N. The Heart: Its History, Its Symbolism, Its Iconography and Its Diseases. Page 9.
2. Boyadjian N. The Heart: Its History, Its Symbolism, Its Iconography and Its Diseases. Page 48.
3. Boyadjian N. The Heart: Its History, Its Symbolism, Its Iconography and Its Diseases. Page 15–20.
4. Bowman IA. The symbolism of the heart: a review. Tex Heart Inst J 1987;14: 337.
5. Boyadjian N. The Heart: Its History, Its Symbolism, Its Iconography and Its Diseases. Page 37–45.
6. Albert R. Jonsen, The Birth of Bioethics (New York: Oxford University Press, 1998) Page 235.
7. Nelson GD. A brief history of cardiac pacing. Texas Heart Inst J 1993;20:12–8, p 12.
8. Ward C. Henderson S. Metcalfe N.H. A short history on pacemakers International Journal of Cardiology, 169 (4), p. 244–248.
9. Sonnenburg D. One good pulse leads to another: Pacemaker technology and how it grew. Encounter 1983;6:6–7.
10. Nelson GD. A brief history of cardiac pacing. Texas Heart Inst J 1993;20:12–8, p 13. Madhavan M, Mulpuru SK, McLeod CJ, et al. Cardiac pacemakers: function, troubleshooting, and management. J Am Coll Cardiol. 2017;69, p 189.
11. Beck H, Boden WE, Patibandla S, et al. 50th anniversary of the first successful permanent pacemaker implantation in the United States: historical review and future directions. Am J Cardiol. 2010;106:812.
12. Kusumoto FM, Goldschlanger N. Device therapy for cardiac arrhythmias. JAMA. 2001; 287:1848–1852.
13. Cazeau, S., Leclercq, C., Lavergne, T., Walker, S., Varma, C., Linde, C., Garrigue, S., Kappenberger, L., Haywood, G.A., Santini, M., Bailleul, C., Daubert, J.C., and Multisite Stimulation in Cardiomyopathies (MUSTIC) Study Investigators. Effects of multisite biventricular pacing in patients with heart failure and intraventricular conduction delay. *N Engl J Med*. 2001; 344: 873–880. Bristow, M.R., Saxon, L.A., Boehmer, J., Krueger, S., Kass, D.A., De Marco, T., Carson, P., DiCarlo, L., DeMets, D., White, B.G., DeVries, D.W., Feldman, A.M., and Comparison of Medical Therapy, Pacing, and Defibrillation in Heart Failure (COMPANION) Investigators. Cardiac-resynchronization therapy with or without an implantable defibrillator in advanced chronic heart failure. *N Engl J Med*. 2004; 350: 2140–2150. Cleland, J.G., Daubert, J.C., Erdmann, E., Freemantle, N., Gras, D., Kappenberger, L., Tavazzi, L., and Cardiac Resynchronization-Heart Failure (CARE-HF) Study Investigators. The effect of cardiac resynchronization on morbidity and mortality in heart failure. *N Engl J Med*. 2005; 352: 1539–1549.
14. Reynolds D, Duray GZ, Omar R, Soejima K, Neuzil P, Zhang S, Narasimhan C, Steinwender C, Brugada J, Lloyd M, Roberts PR, Sagi V, Hummel J, Bongiorni MG, Knops RE, Ellis CR, Gornick CC, Bernabei MA, Laager V, Stromberg K, Williams ER, Hudnall JH, Ritter P; Micra Transcatheter Pacing Study Group. A leadless intracardiac transcatheter pacing system. N Engl J Med. 2016;*374*:533–541.
15. Goldstein NE, Lampert R, Bradley E, Lynn J, Krumholz HM. Management of implantable cardioverter defibrillators in end-of-life care. Ann Intern Med 2004; 141(11):835–838. Kusumoto FM, Goldschlager N. Device therapy for cardiac arrhythmias. JAMA 2002;287(14):1848–1852.
16. Bristow MR, Saxon LA, Boehmer J, et al. Cardiac-resynchronization therapy with or without an implantable defibrillator in advanced chronic heart failure. N Engl J Med 2004;350(21):2140–2150.
17. England R, England T, Coggon J. The ethical and legal implications of deactivating an implantable cardioverter-defibrillator in a patient with terminal cancer. J Med Ethics. 2007 Sep;33(9):538–40. P 2.
18. Grubb BP[1], Karabin B. Ethical Dilemmas and End-of-Life Choices for Patients with Implantable Cardiac Devices: Decisions Regarding Discontinuation of Therapy. Curr Treat Options Cardiovasc Med. 2011 Oct;13(5):385–92. P 386.

19. Kramer DB, Kesselheim AS, Brock DW, Maisel WH. Ethical and legal views of physicians regarding deactivation of cardiac implantable electrical devices: a quantitative assessment. Heart Rhythm. 2010;7(11):1537–42, p 2.

20. Mueller P.S., Jenkins S.M., Bramstedt K.A., Hayes D.L. 2008. Deactivating implanted cardiac devices in terminally ill patients: Practices and attitudes. Pacing Clinical Electrophysiology 31:560–8.

21. Kapa S., Mueller P.S., Hayes D.L., Asirvatham S.J. 2010. Perspectives on withdrawing pacemaker and implantable cardioverter-defibrillator therapies at end of life: Results of a survey of medical and legal professionals and patients. Mayo Clinic Proceedings 85:981–90.

22. Kramer DB, Kesselheim AS, Brock DW, Maisel WH. Ethical and legal views of physicians regarding deactivation of cardiac implantable electrical devices: a quantitative assessment. Heart Rhythm. 2010;7(11):1537–42, Mueller P.S., Jenkins S.M., Bramstedt K.A., Hayes D.L. 2008. Deactivating implanted cardiac devices in terminally ill patients: Practices and attitudes. Pacing Clinical Electrophysiology 31:560–8.

23. Torke AM, Siegler M, Abalos A, Moloney RM, Alexander GC. Physicians' experience with surrogate decision making for hospitalized adults. J Gen Intern Med. 2009;24(9):1023–1028. Silveira MJ, Kim SY, Langa KM. Advance directives and outcomes of surrogate decision making before death. N Engl J Med. 2010;362(13):1211–1218.

24. Goldstein NE, Mehta D, Siddiqui S, et al. Sean Morrison RS. That's like an act of suicide: patients' attitudes toward deactivation of implantable defibrillators. J Gen Intern Med https://doi.org/10.1007/S11606-007-0239-8.

25. Ahmad M, Bloomstein L, Roelke M, et al. Patients' attitudes toward implanted defibrillator shocks. Pacing and Clinical Electrophysiology. 200;23:p 937.

26. Kirkpatrick JN, Gottlieb M, Sehgal P, Patel R, Verdino RJ. Deactivation of implantable cardioverter defibrillators in terminal illness and end of life care. Am J Cardiol. 2012;109:91–94.

27. Marinskis G, van Erven L; EHRA Scientific Initiatives Committee. Deactivation of implanted cardioverter-defibrillators at the end of life: results of the EHRA survey. Europace 2010;12:1176 –1177. Kelley A, Mehta S, Reid M. Management of patients with ICDs at the end of life (EOL): A qualitative study. Am J Hosp Palliat Med 2009;25:440–446.

28. Sulmasy, D.P. 2007. Within you/without you: Biotechnology, ontology, and ethics. Journal of General Internal Medicine 23(Suppl 1): p 70.

29. Goldstein, N.E., D. Mehta, E. Teitlbaum, E.H. Bradley, and R.S. Morrison. 2007. "Its like crossing a bridge" complexities preventing physicians from discussing deactivation of implantable defibrillators at the end of life. *Journal of General Internal Medicine* 23(Suppl. 1): 2–6.

30. Sulmasy, D.P. 207. Within you/without you: Biotechnology, ontology, and ethics. Journal of General Internal Medicine 23(Suppl 1): p 71. Huddle TS, Amos Bailey F. Pacemaker deactivation: withdrawal of support or active ending of life? Theor Med Bioeth. 2012;33(6):p 426.

31. Sulmasy, D.P. 2007. Within you/without you: Biotechnology, ontology, and ethics. Journal of General Internal Medicine 23(Suppl 1): p 71–72.

32. Jansen LA. Hastening death and the boundaries of the self. Bioethics. 2006;20:105–11.

33. Kay, G.N., and G.T. Bittner. 2009. Should implantable cardioverter-defibrillators and permanent pacemakers in patients with terminal illness be deactivated? An ethical distinction. *Circulation Arrhythmia and Electrophysiology* 2: 336–339.

34. Wu EB. The ethics of implantable devices. J Med Ethics. 2007;33:532–533. Paola FA, Walker RM. Deactivating the implantable cardioverter-defibrillator: a biofixture analysis. South Med J. 2000;93:20–3.

35. Kraemer, F. 2011. Ontology or phenomenology? How the LVAD challenges the euthanasia. debate. *Bioethics.* Simon, J.R., and R.L. Fischbach. 2008. Case study: "Doctor, will you turn off my LVAD?". *Hastings Center Report* 38(1): 14–15.

36. Kraemer, F. 2011. Ontology or phenomenology? How the LVAD challenges the euthanasia. debate. *Bioethics.* P 146–148.

37. Merleau-Ponty, Op.Cit.note 65, pp.143; D.W. Smith. Phenomenology. Stanford Encyclopedia of philosophy.
38. Lambert, R.L., D.L. Hayes, G.J. Annas, et al. 2010. HRS expert consensus statement on the management of cardiovascular implantable electronic devices (CIEDs) in patients nearing end of life or requesting withdrawal of therapy. *Heart Rhythm, P 1011*. Grubb BP[1], Karabin B. Ethical Dilemmas and End-of-Life Choices for Patients with Implantable Cardiac Devices: Decisions Regarding Discontinuation of Therapy. Curr Treat Options Cardiovasc Med. 2011 Oct;13(5):385–92. P 386.
39. David F Kelly, Gerard Magill, Henk ten Have, Contemporary Catholic Health Care Ethics, second edition, p 132.
40. Here I used a similar argument provided by Dr. Jeremy Simon on total artificial heart, Simon, J.R., and R.L. Fischbach. 2008. Case study: "Doctor, will you turn off my LVAD?". *Hastings Center Report* 38(1): 14–15.
41. Kraemer, Felicitas. "Ontology or phenomenology? How the LVAD challenges the euthanasia debate." *Bioethics* 27, no. 3 (2013): 140–150. Page 145.
42. Simon, Jeremy R., and Ruth L. Fischbach. "Doctor, will you turn off my LVAD?." *Hastings Center Report* 38, no. 1 (2008): 14–15.
43. Lambert, R.L., D.L. Hayes, G.J. Annas, et al. 2010. HRS expert consensus statement on the management of cardiovascular implantable electronic devices (CIEDs) in patients nearing end of life or requesting withdrawal of therapy. *Heart Rhythm, P 1009*.
44. AMA 1996. AMA Code of Medical Ethics: Policy on End of Life Care: Opinion E-2.20. Chicago, IL: American Medical Association, 1996.
45. Jeremy Sugarman and Daniel P. Sulmasy, ed. Methods in Medical Ethics (Washington, DC: Georgetown University Press, 2010) page 114.
46. Tom, Beauchamp, James F. Childress, Principles of Biomedical Ethics, Seventh Edition 2012, Page 101.
47. Pellegrino ED, Thomasma DC. The conflict between autonomy and beneficence in medical ethics: proposal for a resolution. J Contemp Health Law Policy. 1987;3:24.
48. Pellegrino ED, Thomasma DC. The conflict between autonomy and beneficence in medical ethics: proposal for a resolution. J Contemp Health Law Policy. 1987;3:24–26.
49. Tom, Beauchamp, James F. Childress, Principles of Biomedical Ethics, Seventh Edition 2012, Page 102.
50. Pellegrino ED, Thomasma DC. The conflict between autonomy and beneficence in medical ethics: proposal for a resolution. J Contemp Health Law Policy. 1987;3:31.
51. Pellegrino ED, Thomasma DC. The conflict between autonomy and beneficence in medical ethics: proposal for a resolution. J Contemp Health Law Policy. 1987;3 29–30.
52. Tom, Beauchamp, James F. Childress, Principles of Biomedical Ethics, Seventh Edition 2012, Page 110, Pellegrino ED, Thomasma DC. The conflict between autonomy and beneficence in medical ethics: proposal for a resolution. J Contemp Health Law Policy. 1987;3:33.
53. Blackhall LJ, Murphy ST, Frank G, et al. Ethnicity and attitudes toward patient autonomy. J Am Med Assoc. 1995;274(10):820–825.
54. Lantos J, Matlock AM, Wendler D. Clinician Integrity and Limits to Patient Autonomy. JAMA. 2011;305:498.
55. Aghabarary M, Dehghan N. Medical futility and its challenges: a review study. J Med Ethics Hist Med. 2016;9: page 4.
56. Aghabarary M, Dehghan N. Medical futility and its challenges: a review study. J Med Ethics Hist Med. 2016;9: page 7.
57. Schneiderman L. (2011) Defining medical futility and improving medical care. J Bioeth Inq 8: 126.
58. Schneiderman L. (2011) Defining medical futility and improving medical care. J Bioeth Inq 8: 125.
59. Benjamin MM, Sorkness CA. Practical and ethical considerations in the management of pacemaker and implantable cardiac defibrillator devices in terminally ill patients. Proc (Bayl Univ Med Cent) 2017; 30: 157–60.

60. Ambrus J.L., Ambrus C.M., Mink I.B., Pickren J.W. Causes of death in cancer patients. J. Med. 1975;6:61–64. Klastersky, J., Daneau, D., and Verhest, A.: Causes of death in patients with cancer. Eur. J. Cancer 8:149–154, 197.

Bibliography

1. Aghabarary, Maryam, and Nahid Dehghan Nayeri. 2016. Medical futility and its challenges: A review study. *Journal of Medical Ethics and History of Medicine*.
2. Ahmad, Maha, Lauren Bloomstein, Marc Roelke, et al. 2000. Patients' attitudes toward implanted defibrillator shocks. *Pacing and Clinical Electrophysiology* 23:934–938.
3. AMA 1996. AMA Code of Medical Ethics: Policy on End of Life Care: Opinion E-2.20. Chicago, IL: American Medical Association, 1996.
4. Beauchamp, Tom L., and James F. Childress. 2013. *Principles of biomedical ethics*, 7th edition. Oxford: Oxford University Press.
5. Benjamin, Mina M., and Christine A. Sorkness. 2017. Practical and ethical considerations in the management of pacemaker and implantable cardiac defibrillator devices in terminally ill patients. *Proceedings (Baylor University Medical Center)* 30(2):157–160.
6. Blackhall, Leslie J., Sheila T. Murphy, Gelya Frank, et al. 1995. Ethnicity and attitudes toward patient autonomy. *Journal of the American Medical Association* 274(10):820–825.
7. Buchhalter, Lillian C., Abigale L. Ottenberg, Tracy L. Webster, Keith M. Swetz, David L. Hayes, and Paul S. Mueller. 2014. Features and outcomes of patients who underwent cardiac device deactivation. *JAMA Internal Medicine* 174(1):80–85.
8. Bunzel, Benjamin, Brigitte Schmidl-Mohl, Alice Grundböck, and Gregor Wollenek. 1992. Does changing the heart mean changing personality? A retrospective inquiry on 47 heart transplant patients. *Quality of Life Research* 1(4):251–256.
9. England, Ruth, Tim England, and John Coggon. 2007. The ethical and legal implications of deactivating an implantable cardioverter-defibrillator in a patient with terminal cancer. *Journal of Medical Ethics* 33(9):538–540.
10. Goldstein, Nathan, Elizabeth Bradley, Jessica Zeidman, Davendra Mehta, and R. Sean Morrison. 2009. Barriers to conversations about deactivation of implantable defibrillators in seriously ill patients. *Journal of the American College of Cardiology* 54:371–373.
11. Goldstein, Nathan E., Davendra Mehta, Ezra Teitelbaum, Elizabeth H. Bradley, and R. Sean Morrison. 2007. "Its like crossing a bridge" complexities preventing physicians from discussing deactivation of implantable defibrillators at the end of life. *Journal of General Internal Medicine* 23(Suppl. 1):2–6.
12. Goldstein, Nathan E., Davendra Mehta, Saima Siddiqui, R. Sean Morrison, et al. 2008. That's like an act of suicide: Patients' attitudes toward deactivation of implantable defibrillators. *Journal of General Internal Medicine*. https://doi.org/10.1007/S11606-007-0239-8
13. Grubb, Blair P., and Beverly Karabin. 2011. Ethical dilemmas and end-of-life choices for patients with implantable cardiac devices: decisions regarding discontinuation of therapy. *Current Treatment Options in Cardiovascular Medicine* 13(5):385–392.
14. Huddle, Thomas S., and F. Amos Bailey. 2012. Pacemaker deactivation: Withdrawal of support or active ending of life? *Theoretical Medicine and Bioethics* 33(6):421–433.
15. Hui, David, Zohra Nooruddin, Neha Didwaniya, Rony Dev, Maxine De La Cruz, Sun Hyun Kim, et al. 2014. Concepts and definitions for "actively dying," "end of life," "terminally ill," "terminal care," and "transition of care": A systematic review. *Journal of Pain and Symptom Management*.
16. Hwang, In Cheol, Hong Yup Ahn, Sang Min Park, Jae Yong Shim, and Kyoung Kon Kim. 2013. Clinical changes in terminally ill cancer patients and death within 48 h: When should we refer patients to a separate room? *Support Care Cancer* 21:835–840.
17. Inagaki, Jiro, Victorio Rodriguez, and Gerald P. Bodey. 1974. Causes of death in cancer patients. *Cancer* 33:568.

18. Jansen, Lynn A. 2006. Hastening death and the boundaries of the self. *Bioethics* 20:105–111.
19. Jansen, Lynn A., and Daniel P. Sulmasy. 2002. Proportionality, terminal suffering and the restorative goals of medicine. *Theoretical Medicine and Bioethics* 23:321–337.
20. Jonsen, Albert R. 1998. *The birth of bioethics*. New York: Oxford University Press.
21. Jonsen, Albert R. 2008. *A short history of medical ethics*. New York: Oxford University Press.
22. Kapa, Suraj, Paul S. Mueller, David L. Hayes, and Samuel J. Asirvatham. 2010. Perspectives on withdrawing pacemaker and implantable cardioverter-defibrillator therapies at end of life: Results of a survey of medical and legal professionals and patients. In *Mayo Clinic Proceedings*, vol. 85, pp. 981–990.
23. Kay, G. Neal, and Gregory T. Bittner. 2009. Should implantable cardioverter-defibrillators and permanent pacemakers in patients with terminal illness be deactivated? An ethical distinction. *Circulation Arrhythmia and Electrophysiology* 2:336–339.
24. Kelly, David F., Gerard Magill, and Henk Ten Have. 2013. *Contemporary catholic health care ethics*, 2nd edition. Washington, D.C: Georgetown University Press.
25. Kirkpatrick, James N., Maia Gottlieb, Priya Sehgal, Rutuke Patel, and Ralph J. Verdino. 2012. Deactivation of implantable cardioverter defibrillators in terminal illness and end of life care. *The American Journal of Cardiology* 109:91–94.
26. Klastersky, Jean, Didier Daneau, and Alain Verhest. 1972. Causes of death in patients with cancer. *European Journal of Cancer* 8(2):149–154.
27. Kraemer, Felicitas. 2011. Ontology or phenomenology? How the LVAD challenges the euthanasia debate. *Bioethics*. https://doi.org/10.1111/j.1467-8519.2011.01900.x. Accessed 27 Oct 2011.
28. Kramer, Daniel B., Aaron S. Kesselheim, Lisa Salberg, Dan W. Brock, and William H. Maisel. 2011. Ethical and legal views regarding deactivation of cardiac implantable electrical devices in patients with hypertrophic cardiomyopathy. *The American Journal of Cardiology* 1071–1075. e1075.
29. Lampert, Rachel, David L. Hayes, George J. Annas, et al. 2010. HRS expert consensus statement on the management of cardiovascular implantable electronic devices (CIEDs) in patients nearing end of life or requesting withdrawal of therapy. *Heart Rhythm* 7:1008–1026.
30. Lantos, John, Ann Marie Matlock, and David Wendler. 2011. Clinician integrity and limits to patient autonomy. *JAMA* 305:498.
31. Mackler, Aaron L. 2003. *Introduction to Jewish and catholic bioethics: A comparative analysis*. Washington, D.C: Georgetown University Press.
32. Magill, Gerard. 2017. Using the imagination in normative moral reasoning around the principle of double effect to foster doctrinal development in catholic bioethics. In: *Moral normativity*, ed, J. Gielen. Springer.
33. Mueller, Paul S., C. Christopher Hook, and David L. Hayes. 2003. Ethical analysis of withdrawal of pacemaker or implantable cardioverter-defibrillator support at the end of life. In *Mayo Clinic Proceedings*, vol. 78, pp. 959–963.
34. Mueller, Paul S., C. Christopher Hook, and David L. Hayes. 2003. Ethical analysis of withdrawal of pacemaker or implantable cardioverter-defibrillator support at the end of life. In *Mayo Clinic Proceedings*, vol. 78, no. 8, pp. 959–963.
35. Mueller, Paul S., Sarah M. Jenkins, Katrina A. Bramstedt, and David L. Hayes. 2008. Deactivating implanted cardiac devices in terminally ill patients: Practices and attitudes. *Pacing Clinical Electrophysiology* 31:560–568.
36. Mulpuru, Siva K., Malini Madhavan, Christopher J. McLeod, et al. 2017. Cardiac pacemakers: Function, troubleshooting, and management. *Journal of the American College of Cardiology* 69:189–210.
37. Nelson, G. D. 1993. A brief history of cardiac pacing. *Texas Heart Institute Journal* 20:12–18.
38. Paola, Frederick A., and Robert M. Walker. 2000. Deactivating the implantable cardioverter-defibrillator: A biofixture analysis. *Southern Medical Journal* 93:20–23.
39. Pasalic, Dario, Halena M. Gazelka, Rachel J. Topazian, Lillian C. Buchhalter, Abigale L. Ottenberg, Tracy L. Webster, Keith M. Swetz, and Paul S. Mueller. 2016. Palliative care consultation and associated end-of-life care after pacemaker or implantable cardioverter-defibrillator deactivation. *American Journal of Hospice and Palliative Medicine* 33(10):966–971.

40. Pellegrino, Edmund D. 2000. Decisions to withdraw life-sustaining treatment: A moral algorithm. *JAMA* 283:1065–1067.
41. Pellegrino, Edmund D. 2000. Decisions to withdraw life-sustaining treatment: A moral algorithm. *JAMA* 283(8):1065–1067.
42. Pellegrino, Edmund D., and David C. Thomasma. 1987. The conflict between autonomy and beneficence in medical ethics: Proposal for a resolution. *The Journal of Contemporary Health Law and Policy.*
43. Schneiderman, Lawrence. 2011. Defining medical futility and improving medical care. *Journal of Bioethical Inquiry.*
44. Simon, Jeremy R., and Ruth L. Fischbach. 2008. Case study: "Doctor, will you turn off my LVAD?" *Hastings Center Report* 38(1):14–15.
45. Sugarman, Jeremy, and Daniel P. Sulmasy, ed. 2010. *Methods in medical ethics.* Washington, DC: Georgetown University Press.
46. Sulmasy, Daniel P. 2007. Within you/without you: Biotechnology, ontology, and ethics. *Journal of General Internal Medicine* 23(Suppl 1):69–72.
47. Sulmasy, Daniel P. 1998. Killing and allowing to die: Another look. *Journal of Law, Medicine & Ethics* 26:55–64.
48. Ward, Catherine, Susannah Henderson, and Neil H. Metcalfe. 2013. A short history on pacemakers. *International Journal of Cardiology* 169(4):244–248.
49. Wu, Eugene B. 2007. The ethics of implantable devices. *Journal of Medical Ethics* 33:532–533.

Part IV
End of Life Issues

Chapter 7
Ethical Dilemmas at the End of Life
Brain Death and Palliative Sedation

Questions

- Should we equate brain death with cardiovascular death? And if so, on what scientific grounds?
- What are the differences between palliative sedation and euthanasia?
- Does administration of a muscular blocking agent to paralyze the diaphragm in end the stage of life fulfils the principle of double effect?

7.1 Introduction

This chapter is the first of two chapters that will focus on some ethical dilemmas at the end of life. Delivering care at these unique times and within these unique circumstances can be quite uncomfortable to many health care professionals. Especially with the current fragmentation of health care, connecting with a suffering patient or her family is not by any means an easy task. Brain death is one of those clinical scenarios that can be quite stressful to both medical teams and the patient's family. Deciding to discontinue life support on a beloved person while not being certain about their future prognosis is not an uncommon situation. What are the criteria to declare someone brain dead? What are the theoretical background and the historical narrative of the definition? And how did we end up accepting brain death as a death criterion? And was that acceptance universal? How should we respond to those patients who believe that a miracle could happen to their ill family member? All those questions are quite common, and we, as health care professionals, encounter them on a daily basis. Unfortunately, many of those who end up taking care of terminally ill patients are not trained in delivering care at the end of life. Many studies showed a lack of knowledge in the basic principles of pain control, sedation, and symptom control in this patient population. Moreover, there is still significant confusion regarding the theoretical differences between the concepts of palliative sedation, sedation to unconsciousness, euthanasia, and physician-assisted suicide. This lack of knowledge about the ethical differences between those notions complicates the decision-making process and creates significant anxiety for patients, families, and health care teams. These two chapters aim to examine some of those dilemmas at the end of life, to clarify the confusion around the common concepts, and hopefully help those in charge in delivering better care to their suffering patients.

7.2   Brain Death

7.2.1   Historical Review

Humans have always been perplexed by the mystery of death. In fact, it is pretty uncommon to study any civilization without coming across its encounter with end of life. Despite the differences in their beliefs, these civilizations shared many intuitions about the mystery of death, such as the sanctity of the departure moment, the soul's reincarnation, and the dichotomized nature of the afterlife. Throughout history, man tried desperately to overcome this inevitable destiny. He practiced medicine, wrote narratives and legends, and tried to influence that transcendent realm through prayers and rituals.[1]

For a long time, the cessation of breathing was considered alone a clear sign of death. Interestingly, in many ancient languages, "soul" and "breath" share the same word; for example, in Aramaic "Nafsho" and in Hebrew, "Nafash," both mean breath.[2] This theoretical observation was later grounded on a better understanding of the complex physiology of the Cardio-pulmonary circuit. The physicians of the sixteenth century noticed that any interruption in this blood flow to the organs would eventually lead to a decrease in oxygen supply to the brain, which will subsequently be followed by coma and death.[3]

In the middle of the twentieth century, the previous recognition of death was unexpectedly challenged by the introduction of positive pressure ventilation.[4] During the polio pandemic, physicians started to recognize some cases in which many patients stayed in a deep-unconscious state despite surviving the acute polio infection. The patients were unresponsive and unable to breathe without vent support. A group of French neurologists named this clinical state Coma Depasse or Ultra Coma.[5] Puzzled by this challenge, the leaders in the field of Anesthesiology wrote to Pope Pius XII seeking further moral guidance on how should the medical society treat those patients. Should these hopelessly unconscious patients be considered dead? Are physicians obliged to continue to use modern artificial respiration equipment to keep these unfortunate patients alive?

The Pope answered the previous question by emphasizing two points; the first discussed the concept of extraordinary means of care, which he stated, could be withdrawn if the decision was congruent with the patient's and family's wishes. The second point emphasized that determining the moment of death is a biological fact that falls within the medical-practice capacity and not the church authority. The

Pope invited the physicians to re-define the concept of death in the light of these new technological advancements.[6]

During the same period, scientific progress brought to the table another challenging dilemma, namely, organ transplantation. Whether by chance or by destiny, these two paths (Brain death and Organ Transplantation) crossed each other.[7] In 1966, the Ciba Foundation Conference on organ transplantation discussed the necessity of refining the definition of death and the need to use brain-injured patients, or as they called them, "living cadavers" to provide more viable kidneys for those in need.[8] This Utilitarian inclination was further empowered by the pioneering success of the first heart transplantation in December 1967. Following this achievement, the Harvard Committee was formed and started working on its famous report on brain death, which was later published in August of 1968.[9] In this report, the clinical criteria for determining brain death included the following elements, (1) Coma: Unreceptivity and Unresponsitivity, (2) No movements or Breathing, (3) No brain stem and peripheral reflexes, and (4) flat EEG.[10] Ironically, the Harvard report physicians cited no evidence to their clinical criteria beyond the Pope Pius XII speech.[11] In the following years, many states started accepting brain death as equivalent to ordinary death. On July 9th, 1981, the President's Commission released a comprehensive report: The Uniform Determination of Death Act. According to this report: A person is considered dead if he sustained either (1) irreversible cessation of cardiorespiratory function or (2) irreversible cessation of the function of the entire brain, including the brain stem.[12]

Despite its wide adoption, brain death remains a pretty controversial topic. As we will see later, the bio-philosophical grounding of the concept lacks substantial evidence, and empirical data regarding the patient's diagnosis and prognosis raises many concerns about the validity of the adopted criteria.

7.2.2 From Physics to Metaphysics

Before discussing the different views regarding brain death, we have to elaborate on two essential topics: (1) the second law of thermodynamics, known as the law of entropy, (2) the concept of the human Soul.

Entropy is defined as the measure of disorganization.[13] It states that energy tends to disperse and move into a more disorganized state in a closed system. This definition is quite confusing and not easy to grasp. In simple language, the odds that a given number of molecules will set themselves in an organized fashion are much lower than the odds of being randomly and chaotically distributed if things are left to occur spontaneously without external force. Take the following example, if you took a box of puzzle pieces and threw it in the room, what are the odds that all the pieces will fall and perfectly align to create the puzzle picture? Unimaginably low, close to impossible. In a philosophical sense, all things in the universe tend to move to a disorganized state. If you take a heated metal bar and put it in a cold room, the heat will disperse, and the metal bar temperature will equilibrate with the room temperature. Now think about the human body. We, like all other living organisms, are open systems. We exchange energy and matter with the surrounding environment in an attempt to resist entropy. Our life is a constant work of overcoming entropy. We do not lose our heat and equilibrate with room temperature because we constantly produce heat by metabolism. It is like putting a coffee cup on a heating pad; it will stay warm because it is no longer in an isolated state (closed system). Likewise, to stay in this hemostatic state, our body needs food and metabolism instead of electricity and a heating pad in the coffee cup example. (13)

The body also needs to process and react to the information inside and outside itself. We receive inputs from the environment, and we process them. When we see a tiger, we resist entropy by running away. When we notice a shiny green apple, we eat it to get calories. Because our central nervous system is remarkably complicated, we can also process environmental stimuli in an extremely sophisticated fashion. Take the following example, the sentence: "Stop!!! There is a cliff", might mean nothing to a group of E.coli, might be a sound vibration for the worm, but might save someone's life on a hiking trip!!! Now how does all this align with brain death?[14] The main theory that drove the acceptance of the definition of brain death as an equal state to ordinary death was the role of the brain as a central organ of integration. Without our brain, the human body will lose its capacity to resist entropy and survive the complex environment in which we live. The brain is the organ responsible for maintaining all the necessary biological functions, breathing, circulation, heat, and volume hemostasis; without its central integrating capacity, we cannot survive.

The second essential topic to elaborate on is the concept of the human soul. I will be referring here to three different metaphysical positions. The first is materialistic monism, which refuses the human soul's presence in its metaphysical dimension and reduces the human being, including his intellectual capacity, to his brain. In simple words, if the brain is destroyed, the person ceases to exist.[15]

The second is Mind/Body dualism, held by Plato and Descartes, in which both acknowledged the presence of a metaphysical soul not reduced to matter but thought that the body and soul are two separate components. For them, the body is a machine, and the mind is mainly a spiritual form responsible for moving the body.[16] The third position is Hylomorphic, held initially by Aristotle and adopted later by Aquinas. It is the official position of the Catholic church regarding the nature of the body and soul.[17] According to this position, a human being is a hybrid spiritual/physical being. His body is enspirited, and his soul is enfleshed.[18] The soul makes the human body a human body, and it is what holds it together. It is the substantial form and the life-principle of the human being. The soul is not located in one organ and is essentially the central integrator.[19] Without it, the human body ceases to exist and becomes a corpus. As we will be discussing later, this view contradicts the assumption that the brain as an organ is the central body integrator and the main force behind a living body's ability to resist entropy. As Shewmon argued, a brain-dead person is still a human being with paralyzed mental power. He is still a human, and the loss of brain function does not annihilate him, as cutting a piano's strings does not make a musician any less of a pianist.[20] Before discussing the different arguments around brain death, it is worth adding another point to the discussion of human soul. Aquinas thought that the human soul has an intellectual capacity that differentiates it from animals and plants. Animals, according to him, had two souls, sentient and nutritive, whereas plants had only one soul, the nutritive soul. In both cases, he thoughts that plants and animals could not exist outside this hybrid nature. (19)

7.3 Definitions and Disagreements

Anatomically, the human brain is divided into three main parts. (1) The Brain stem, which contains the control centers of breathing and circulation. (2) The Diencephalon, which contains the Thalamus and Hypothalamus and is responsible for thermoregulation, appetite, and neuroendocrine control. (3) The Brain Cortex which is responsible for higher brain functions, such as, fine motor, fine sensory and intellectual abilities.

Another essential concept regarding brain death is Consciousness. It remains, so far, the most fascinating mystery of our existence. Not fully understood; however, it is generally discussed in two main spheres: awakefulness and awareness. The first is mainly attained by the function of the reticular formation in brain stem, and the second is primarily a result of the function of the cortex and Thalamo-cortical and Cortico-Cortical connections.[21] Now, let's start closely examining this dilemma.

7.3.1 Brain Death Equals Death

The legislation of the Uniform Determination of Death was greatly influenced by the work of James Bernat. In March 1981, Bernat published a prominent article supporting equating brain death with cardiovascular death. He grounded his theory on the presumption that the human brain is the primary organ responsible for integrating the organism's function as a whole. Bernat started by making a fundamental distinction between the definition of death, the criteria for determining death, and the tests needed to confirm the criteria.[22] For him, death was an event that separated the process of dying from the process of disintegration.[23] He defined death as the permanent cessation of the organism as a whole. Death was a biological phenomenon shared between animals and humans, and it should not be confused with the human's loss of personhood and ability to carry on highly interactive and intellectual functions. Therefore, patients with PVS are not dead, although they might have lost what is essentially significant to express human nature. Bernard hypothesized that the human brain, through its three anatomical parts: the brain stem, the cortex, and the neuroendocrine function, was the central integrator of the organism as a whole. Without the brain, the organism is unable to maintain its integration and resist entropy.[24] He saw that classical cardiovascular-respiratory death criteria as a part of the big picture of brain death. Because eventually, the lack of blood supply to the brain would lead to brain anoxia and the irreversible loss of the organ's integration. In the same article, he discussed the pathophysiology of death by Hanging. His explanation was grounded on the presumption that anoxic brain injury following apnea (which results from brain stem compression after odontoid process fracture) is the main reason for death. Because the insult would eventually be followed by cessation of the cardiac function since the human heart cannot continue to pump, for a long time, after losing the neurological support of the brain stem.

Over the following few decades, and in the face of many criticisms, Bernat provided further points to support his brain death argument. In 2006, he published an article titled, "The Whole-Brain Death Concept Remains Optimum Public Policy." In this article, he acknowledged the valid criticism of his theory of the brain's central role in the organism's integration. However, he defended his position by arguing that the biological concept of death should be examined through the lens of practical public policy.[25] He argued that we should balance the minor limitations of brain death definition with the goodness achieved through organ transplantation. He stated that the five-minute role of non-heart-beating organ donation also has its limitation because physicians might still be able to resuscitate dead patients after five minutes of death.

Recently, Bernat published another work defending brain death, focusing on the importance of consciousness as the highest property and capacity of a human being. Bernat grounded his argument on Mereology and theoretical biology.[26] He stated that the mereological characteristics of an organism have a teleological aspect pointing toward its survival. In layman's terms, the various organs of a human interact to protect the organism from entropy and preserve survivability. This mereological property is mainly supported by the human being's highest emergent property, which is consciousness. For Bernat, loss of consciousness is necessary for declaring death, although it is not sufficient.[27] Bernat argued that consciousness is the only function in the human body that cannot be replaced by technology, and if we were able one day to introduce consciousness by transplanting brains, then we have to rethink the definition of life and death. Bernat's approach here is very close to identifying human life through higher brain functionality and personhood, although he is grounding his argument on the brain's teleological role in maintaining the integration of the organism and promoting its survival.

7.3.2   Brain Death is not Equal to Death

The controversy regarding the concept of brain death can also be viewed under three main categories: definition, criteria, and testing. Alan Shewmon, one of the leading scholars in examining the concept of brain death, pointed out that there are three main views to define brain death. The first is sociological relativistic. This approach considers death to be a socially constructed notion where society, based on its vision, decides when to consider a person dead.[28] Henry Beecher, the chairman of the Harvard Committee, commented on the famous Harvard brain death report stating that society has to choose a state to define death where it can benefit from harvesting the organs of hopelessly unconscious patients.[29] Shewmon rejected this approach of defining death because it is not grounded on a solid metaphysics of life and death. Shewmon criticized another approach in defining brain death, namely, the person/mind reductionist one. This approach takes the Platonic/Cartesian view, which primarily emphasizes the importance of the cognitive capacity of the human being, "cogito ergo sum."[30] According to this view, the soul is the conscious mind, and the body is a mere mechanical machine. This approach is the mainstream for the majority of contemporary scientists. However, the mind/soul is now merely reduced to the function of the brain tissue. A Person/Brain reductionism, which equates the lack of existential consciousness of a human being with his death. The death of what is necessary to be human. Nevertheless, this approach could also be refused because death is a mere biological phenomenon that should be shared between all vertebral animals, including human beings. Moreover, it is challenging to justify considering a patient in a PVS state to be dead. Permanent loss of consciousness is not equal to death.[31]

The third view in defining death is mainly biological, which is the loss of the integrative unity of the body as a whole. According to Shewmon, the metaphysical background of this view is reliable and also compatible with the Judeo-Christian tradition. The soul, which is the life-principle of a human being, is the central integrator, and death, in the metaphysical sense, is a separation between the soul and the body. Shewmon refused to accept the brain as a central dictator over other body functions. He provided many examples of how the bodies of brain-dead patients continue to perform many holistic functions that cannot be reduced to their components. For instance, homeostasis, assimilation of nutrients, and sexual maturation.[32] Moreover, one of the main rationales for equating brain death with ordinary death is the assumption that after losing brain input, cardiovascular-respiratory collapse is inevitable. Shewmon argued that first, this assumption mixes up diagnosis with prognosis. Second, the disintegration and eminent loss of cardio-respiratory function following brain death are majorly influenced by non-brain related factors.[33] Shewmon published a Meta-analysis of 175 cases of brain death patients in which he showed that survival capacity correlated inversely with age and patients with primary brain injuries had longer survival compared to patients with systemic syndromes. Also, not all patients died shortly after confirming the diagnosis of brain death. In the same Meta-analysis, survival varied, where some patients continued to live for weeks while others continued to survive even for years post the diagnosis of brain death.[34]

The second inconsistency regarding the concept of brain death is related to the clinical criteria and tests. The American Academy of Neurology published a practice guideline in 1995 for determining brain death. The guidelines were updated in 2010 and include the three following clinical findings to confirm brain death: Coma, absence of brain stem reflexes, and apnea test.[35] The theoretical rationale is the cessation of the function of the entire brain-the brain as a whole. Those criteria are the same ones adopted by the Harvard Committee three decades before. The guidelines emphasize the necessity of lacking eye-opening and all signs of responsiveness.[36] The brain stem reflexes should include pupillary response, corneal reflexes, phalangeal and tracheal reflexes. And finally, the apnea test to rule out the presence of any spontaneous breathing once the CO_2 level reached 60 mm Hg. The guidelines don't determine the observation period for neurological recovery nor mention any evidence of adopting the previously mentioned clinical signs.[37] Indeed, the theoretical problems with this brain death criteria are that they do not test all brain functions. For example, the hypothalamus/pituitary function and water and heat regulation are not part of the requirements needed to determine brain death. Even the updated version of the AAN position in 2019 clearly states that despite being aware of the preservation of neuroendocrine function, the AAN still considers this finding not inconsistent with the brain death definition.[38]

Furthermore, The AAN states that since the publication of their guidelines in 1995 for determining brain death, there have been no case reports of neurological recovery to invalidate the current criteria. Many scholars reject this statement. Repertinger published a case report of a boy who sustained a massive brain injury at the age of four from bacterial meningitis.[39] The child lived for two decades post that injury. He met all the brain death criteria, and his body did not crumble or lack integration functions. The autopsy confirmed the presence of no viable neurological tissue in his calcified brain. Another case is Jahi Mcmath, a 13-year-old female who developed post-op cardiac arrest and was later diagnosed with brain death secondary to anoxic brain injury. The patient's family refused to withdraw care and requested a hospital transfer to New Jersey, which is the only state that allows the refusal of brain death. Jahi lived for five years post-primary injury and died in 2018, having two death certificates, one from California in 2013 and one from New Jersey in 2018.[40] Shewmon argued that many brain death patients who can survive the acute phase of autonomic dysfunction and spinal shock that follow brain death might have better outcomes. Unfortunately, we do not encounter many cases because most families withdraw care, or the patient is declared organ donors.[41]

The ancillary tests for determining brain death also have major problems. For example, the AAN report states that brain death is a clinical diagnosis, and the lack of EEG activity is not necessary for confirming that.[42] In that case, the AAN put itself in the position of brain stem death and not the cessation of whole-brain function.[43] Also, the lack of EEG activity might be in the setting of global ischemic penumbra where the total blood supply to the brain decreases in the setting of acute injury and might give a false-negative impression of lack of cortical activity.[44] Again, what about those who meet the criteria for brain death but have some EEG activity? Some scholars went on to say that this might indicate a neurological activity but not a meaningful function.[45] But how do we know? The evidence surrounding all these ancillary testing is quite weak (Level U of evidence).[46] The same pathophysiology of ischemic penumbra can explain the lack of cerebral blood flow on radionuclide angiography. The cerebral blood flow will decrease to a point where the preservation of neurons is maintained; however, the flow is below the test detection threshold. Furthermore, postmortem studies showed that 60% of brain autopsies of heart beating donors did not have major structural changes (44). Apnea testing can also have detrimental effects on the brain. Nguyen pointed out that despite making sure to preserve adequate oxygenation during the test, the rise of CO_2 levels can cause significant vasodilatation in the brain vessels, which can increase intracranial pressure and further worsen ischemic injury.[47]

Lastly, proponents of brain death always cite the argument of physiological decapitation to support the conceptualization of brain death. Alan Shewmon provided a detailed critique of this analogy. The separation of the head from the body during decapitation provokes a horrified emotional trauma in the observer. However, the lack of a central brain integration role is not the basis for equating brain death with physiological decapitation. Shewmon argued that what if the guillotine blade stopped 1 mm before totally cutting the neck of a human being, then the vessels were sutured to maintain hemostasis? What would be the cause of death in that case? Shewmon stated that someone could divide the anatomical compartments of the neck into three parts. The soft tissues have no role in the analogy with decapitation. The vessels also are not the leading cause of death in the analogy. Finally, what about those who get high cervical cord injury, and for the sake of argument, let's say that they also had vagotomy and destruction of ADH center in the brain, do we consider those people dead? Those people technically had lost their entire brain input, and the theory of central integration is debunked in that case, given that they are not considered brain dead.[48]

Moreover, evidence surrounding the concept of brain death is quite concerning. In a survey of 218 neurologists in the united states, 27% of the responders chose the loss of integration to justify equating brain death with death, and 48% chose higher brain functions and loss of consciousness, clearly not differentiating the lack of personhood with death.[49] In the same survey, 70% of the responders stated that the presence of EEG activity is not compatible with brain death. Studies also showed wide variability in following the AAN recommendations in determining brain death. For example, a study of 76 patients with brain death at the University of Missouri showed that 76.3% of the cases had a fully documented neurological examination, and only 39.5% had apnea testing done to confirm the diagnosis.[50]

To summarize, brain death continues to be a challenging ethical dilemma. Physicians hold different opinions regarding the concept. Both sides of the debate agree with the definition of death based on the second principle of thermodynamics. However, they disagree on whether the brain is solely responsible for integrating organ functions and resisting entropy. The second major area of controversy is present in the guidelines for diagnosing brain death. Many of the recommended clinical findings and ancillary testing are not based on high-level evidence and, when confirmed, do not fulfill the theoretical grounding of "whole brain" death. Physicians might agree or disagree on whether brain death should be totally equated with cardio-respiratory death. Yet probably, this disagreement has deeper roots, not obviously related to the definitions and tests, but rather to the actions needed to be taken after, as the conversation shifts toward end-of-life care, organ harvesting, palliative sedation, and euthanasia.

7.4   Palliative Sedation

Dealing with a patient's suffering at the end of life can be quite challenging. Physicians and other health care professionals often find themselves struggling with the care of imminently dying patients. Families are usually stressed about their beloved ones, and patients are not uncommonly confused and unable to communicate and express their symptoms. Overall the circumstances at the bedside are bar far not ideal. With the advancement of our understanding of pain physiology, consciousness, and analgesia, palliative sedation emerged as a potential therapy for refractory symptoms at the end of life. The literature around this topic remarkably grew and expanded over the past three decades. In response to that, many medical societies released guidelines and positions to protocolize the practice of palliative sedation. However, similar to many other procedures and therapies in modern medicine, palliative sedation created its own ethical and non-ethical dilemmas. Some of those areas of controversy are the timing of initiating sedation, the medications that should be given, the degree and the extent of sedation, the continuation of artificial feeding, and the differentiation of this procedure from other practices at the end of life, such as physician-assisted suicide and euthanasia.

7.4.1   Definitions and General Facts

In 1997 the U.S. Supreme Court expressed its support for palliative sedation. The procedure was labeled at that time "terminal sedation.", a term that is rarely used at present. Sandra O'Connor, the former Associate Justice of the Supreme Court, stated the following "A patient who is suffering from a terminal illness and who is experiencing great pain has no legal barriers to obtaining medication from qualified physicians, even to the point of causing unconsciousness and hastening death."[51] At present, many definitions of palliative sedation are there in clinical literature, with variable nuances reflecting the differences in their philosophical underpinning. For example, the National Ethics Committee of the Veterans Health Administration (NEC-VHA) defines palliative sedation as "the administration of non-opioid drugs to sedate a terminally ill patient to unconsciousness as an intervention of last resort to treat severe, refractory pain or other clinical symptoms that have not been relieved by aggressive, symptoms-specific palliation."[52] The National Hospice and Palliative Care Organization defines Palliative sedation as "the controlled administration of sedative medications to reduce patient consciousness to the minimum extend necessary to render intolerable and refractory suffering tolerable."[53] Another commonly encountered definition in clinical literature is Proportionate Palliative Sedation (PPS). Proportionality entails that the benefits of any treatment should balance the burdens, and the dose of the medication should be administered proportionately, gradually being increased to the required degree to relieve the symptoms. Timothy Quill suggested another two definitions to clarify the concept further: Ordinary Sedation, which is the sedation used in daily medical practice within and outside palliation, and Palliative Sedation to Unconsciousness (PSU), in which the aim of sedation is to induce unconsciousness intentionally rather than observing unconsciousness as a physiological side effect of symptoms management.[54] Another form of Palliative Sedation, called Continuous Sedation to Death (CSD), is also frequently encountered in the medical literature. It reflects some of the aspects of the practice of palliative sedation in countries where Euthanasia is legal, such as the Netherlands and Belgium. This form of practice was illustrated in the UNBIASED study, which compared CSD in the previously mentioned countries to The United Kingdom.[55] The study revealed differences in the protocols of CSD; while UK health care professionals started with minimal doses of sedatives and increased the titration to achieve patient' comfort and symptom relief, Dutch and Belgian physicians aimed to target unconsciousness rapidly by either starting with higher doses of sedatives or by quickly titrating up the sedative. One of the nurses in this study commented, "If the patient is still here tomorrow, we double the dose; the patient is not awake anymore; what is the point of letting her lay here for days?".[56]

Given this lack of clarity in the definition, Palliative Sedation as a practice continues to cause considerable unease and discomfort to all those involved in a dying person's care. In a review of thirty-three research papers, Claessens and her colleagues showed that up to 25% of families experience significant distress while making decisions to initiate PS for their beloved ones.[57] A survey of nurses in Belgium also showed that health care professionals experienced ethical challenges regarding PS, with 77% of nursing staff stating that continuous deep sedation hastens the death of the patient.[58] Similar emotional and ethical burdens were observed in a qualitative study of seven hospice nurses in the UK. All of them expressed having significant anxiety about the practice of palliative sedation and whether they are facilitating the death of the patients by administering various sedatives.[59] Physicians as well had different opinions regarding the practice of palliative sedation. In the large survey of 1156 physicians in the United States, two-thirds of the physicians disagreed with primarily aiming to induce unconsciousness in palliative sedation; however, they agreed on accepting unconsciousness as a side effect. 59% disagreed that sometimes intentionally hastening death is the right thing to do to a terminally ill patient, and only 5% strongly agreed with initiating palliative sedation to relieve existential (psycho-social suffering).[60]

7.4.2   The Principle of Double Effect

The principle of double effect is one of the most commonly cited principles in medical ethics. Thomas Equines initially articulated it in his work on self-defense in the Summa Theologica. The principle offers moral guidance when there is an action with two foreseen effects, one is bad, and the other is good. According to the rule of double effect, to be morally permissible, the act should fulfill five conditions: (1) first, it must have two effects: one is bad, and the other is good. (2) The act itself should not be intrinsically evil. (3) A person should sincerely intend the good and not the evil consequences of his action. (4) The bad effect should not be the mean of achieving or the cause of the good effect; in other words, the bad effect should be a side effect of the action. (5) The good effect should proportionately be much greater than the bad effect.[61] For example, targeting innocent civilians in the war to weaken the enemy is not morally permissible, whereas targeting military units with the possibility of causing harm to innocent civilians is permissible under the rule of double effect. As long as the person who is committing the action is sincerely trying to avoid or reduce the foreseen but undesired effect, and as long as, the means to achieve that end is proportional to the degree needed to achieve the end. For example, using a laser-guided missile rather than a blind bombardment of military units in a city, as the latter can cause more casualties. The same can be applied in the case of self-defense. A person's interaction with an assailant can lead to the latter's death. As the long the person's violence against the assailant is proportional, meaning it is extending to a degree required to defend himself, and as long as his intentions are focused on protecting himself and not killing perse, his action is justified even if it led at the end to the assailant's death.[62]

The rule of double effect is often cited in the arguments around palliative sedation. Although the principle is often abused and misapplied. For example, the justification of sedating a patient to achieve symptom relief even if the latter can hasten death is not justified under the principle of double effect unless a physician considers sedation (itself) to be a morally good action. Even if the sedative dose is titrated proportionately to the degree needed for symptom relief, it does not mean the action had already fulfilled the double effect rule and hence became a morally justifiable action. Now, I realize that this might be quite confusing, so let me clarify. Take the following scenario: a 78 years old woman is suffering from terminal ovarian cancer and is now being admitted with community-acquired pneumonia and respiratory failure. The patient and her family, after discussing her prognosis with the medical staff, agreed to focus on symptom control and palliative route. The physician in charge decided to start escalating the doses of morphine to control the patient's pains and dyspnea.

Morphine is an archetypal opioid that mainly induces analgesia by activating the MOP receptors in the midbrain.[63] Morphine also relief refractory dyspnea by decreasing respiratory drive, lowering patients' anxiety, and altering the activity of opioids receptors in the lungs.[64] This physiological effect was supported by a recent systematic review of 12 studies which found some evidence supporting the use of morphine for refractory dyspnea.[65] Now morphine can theoretically decrease consciousness level, lead to somnolence and respiratory acidosis, an effect that is foreseen and known to all physicians. The administration of morphine to treat dyspnea/pain is not an intrinsically evil action (second condition of the rule of double effect). The physician, in this case, is not intending to cause the patient's death by administering morphine. He desires to relieve the patient's symptoms, although he foresees that side effects might happen. The physician, as a free moral agent, chooses to act with good intentions. As Aristotle puts it, ethics is about what to do, when what to do is our choice.[66] This intentionality is essential to the application of the rule of double effect, and it differentiates the action of palliative sedation from euthanasia and physician-assisted suicide (third condition). The same analysis can be applied to the administration of benzodiazepines for anxiolysis. If dyspnea did not improve by opioids alone, adding benzodiazepines can lead to a synergistic effect. A randomized controlled trial of (n = 101) patients in Italy confirmed that and showed the midazolam/morphine regimen's superiority compared to morphine and midazolam alone in relieving refractory dyspnea.[67] Now, what about relieving terminal suffering (dyspnea, pain, agitation, restlessness) by aiming to decrease the level of consciousness? Does the rule of double effect have something to say here? For the first impression, someone might think that the answer is yes, but sedating someone with propofol, for example, just for the sake of dissociating him from the surrounding reality, even if it was done proportionally without causing significant respiratory suppression, cannot be justified alone by the rule of double effect. This practice, which is often labeled as proportionate palliative sedation, is not justified by the rule of double effect. The proportionality here is not between two effects like in the case of morphine/benzodiazepines (Pain/dyspnea versus decreased consciousness) but rather in the spectrum of the acceptable degree of lack of consciousness. There is no double effect here, it is a degree of single effect targeted, which is not by itself a bad practice, and it falls under another principle in medicine, which is the parsimonious use of medications.[68]

What about using Rocuronium to relieve a patient's shortness of breath? This question takes us to the application of the fourth condition of the principle of double effect. The bad effect should not be the mean or the cause of the good effect. The administration of a neuromuscular blocking agent will paralyze the respiratory muscles and lead, of course, to decrease dyspnea. However, it will also surely cause the patient's death. In this case, the good effect, which is symptom relief, was achieved directly by the mean. There is a double effect here; however, the bad effect led to the good effect, and the rule cannot justify this. Finally, the fifth condition entails that the good effect of the procedure should be proportionally much greater than the bad effect. When a physician is taking care of terminally ill patients, symptom control and quality of life become the main focus. Relieving suffering in the light of known terminal illness is the priority. Under these conditions, relieving the agonizing dyspnea, pain and restlessness outweigh in their importance the foreseen side effects of decreasing respiratory drive, unconsciousness, and the possibility of hastening death. Although again, to emphasize, consciousness should be preserved when possible for two reasons. First, sedation itself without appropriate symptom control (dyspnea, anxiety, restlessness) can mask the patient's suffering, with some clinical studies showing that it can even increase pain perception.[69] Second, consciousness is necessary for all our unique humane actions such as free choice, aesthetics, love, reasoning, worship, and praying.[70] Dissociating a patient from reality, by sedating him, just for the sake of reducing his ability to communicate and feel, unless being carried out for clear medical indications (for example, lack of sleep for three days or conducting a surgery) is a questionable practice. The biographical (social life) of a terminally ill patient should be preserved if possible. And high doses of sedatives to target consciousness should be left as a last resort only when other options fail. That was the focus of hospice philosophy, which emphasized the importance of connectedness, community, and belonging in the final hours of human life.[71]

In real-time practice, the application of palliative sedation depends to a great extent on the condition of the patient. For example, if the patient was intubated in the ICU, and the code status was changed to DNR-CC (Comfort Care), many clinicians will continue the Fentanyl/Propofol drip after extubating the patient. Others might switch the propofol to midazolam to target anxiety to relieve dyspnea. In medical wards and hospice facilities, the situation might be different. Patients are usually not profoundly sedated with medications, although some degree of natural sedation (delirium, somnolence) almost always occurs.[72] In such cases, the application of palliative sedation might require a different set of expertise and more attention to the importance of preserving biographical life. One of the major issues complicating the practice of palliative sedation is the lack of formal guidelines in the United States. For example, regarding the level of sedation, the ACP-ASIM (American College of Physicians and The American Society of Internal Medicine) suggests a rapid increase in the sedative dose to achieve unconsciousness until death.[73] The NHPCO (National Hospice and Palliative Care Organization) and AAHPM both describe what is most consistent with PPS (proportionate palliative sedation) with gradually increasing the medication dosage to achieve unconsciousness.[74] Another area of inconsistency is the timing of initiating sedation. The NHPCO statement refers to a prognosis of two weeks or less. The AAHPM suggests "at the very end of life," and the ACP describes a prognosis of days to weeks.[75] The positions of professional societies also differ regarding initiating palliative sedation for existential suffering. Existential suffering is the nonphysical suffering that a patient might experience at the end of life. It is, as Sulmasy expressed, the experience of bodily finitude through pain and sickness.[76] The NHPCO statement refers to the lack of consensus regarding initiating palliative sedation for existential suffering. The AAHPM also states that there is a lack of consensus but suggests consulting expert staff in treating mental and spiritual suffering.

7.5   Miscellaneous Points

As I have been discussing so far, taking decisions in patients' care at the end of life can be quite stressful for health care professionals. Not only because of the uncertainty regarding many of the consequences of those decisions but also because the choices directly deal with death, a point where there is no return. The advancements in medical technologies offered so many graces and benefits and saved many patients' lives, yet they opened up a variety of questions regarding whether or not and to what extent we should commit a suffering patient to a technological fate. These advancements challenged our understanding of what is really humane and raised serious questions about the value of human life.

7.5.1 On the Value of Human Life

It is pretty challenging to examine such atopic; in the end, our humanity by itself is probably the greatest mystery. Yet, many of those who work in the health care field and encounter sickness on daily bases have probably thought about this question. Many of them, including myself, believe that human life has a core-intrinsic value regardless of its attributes. It is difficult to comprehend why a physician would stay awake a whole night next to an 89 year old grandfather with dementia on two vasoactive pressors with end-stage cancer and multiorgan failure unless there is something worthy about life itself. This patient, at that state, is unable to attribute anything to his or others' life during those moments of complete dependency. In other words, the value of life could not be only a utilitarian value.[77] However, if we all agree that human life has intrinsic value, the aching question remains: To what extent should it be preserved?

Since the beginning of the scientific revolution in the seventeenth century, the catholic church's teachings emphasized that life is sacred, and a person should not take the life of an innocent man or his own. However, the church's teaching also stressed that life need not be maintained at any cost.[78] This notion was emphasized later by Pope Pius XII in 1958 with the distinction between ordinary and extraordinary treatments. The Catholic Tradition rejected both the extremes of Vitalism and Subjectivism and stated that the life of a human being need not be prolonged in the light of imminent death and when the burdens of treatments outweigh the meaningful benefits achieved by those treatments.[79]

7.5.2 Ordinary and Extraordinary Treatments

The distinction between ordinary and extraordinary treatments is not only a medical one. Although it seems to some extent, obscure and vague, it is still valuable as long as it is taken within the context of a clinical situation. Simple hospital admission of an elderly father to treat community-acquired pneumonia in the United States would be an ordinary treatment that someone should not refuse. However, the same admission might be extraordinary in a developing country during wartime, where the hospitalization price would be a burden on a whole family suffering from hunger, cold, and lack of basic life needs. Keven Wildes reviewed the literature and gathered five elements to designate a treatment extraordinary.[80] These include (1) Impossibility factor: when the means or the usage of the means of treatment are impossible. (2) The means that require extraordinary effort, for example traveling many miles to bring medication. (3) Means of very high expenses. (4) When the means or their use can cause significant and intolerable pain. (5) When the emotions of fear and burdensome outweigh the benefits that would be achieved by applying the treatments. On the other side, an ordinary treatment should offer some meaningful hope for the patient. The absence of hope is sufficient to label a treatment extraordinary, whereas the presence of hope is necessary but not sufficient for a treatment to be ordinary.[81] If the treatment burdens were so profound when they were weighed out by the possible benefits, then the same treatment that was ordinary for a patient might be extraordinary for another. Again, the decision here is not purely scientific. Physicians who practice on a daily basis know that any medical decision involves a nexus of many factors, including scientific knowledge, character virtues, social and moral values.

This philosophical explanation might seem very abstract. Unfortunately, there is no solitary decision algorithm to be followed 100% at the time with guaranteed results in each clinical encounter. There is no escape from some sort of subjectivity regarding the extent to which a patient would accept the treatment's burden. Or the extent to which he will consider the foreseen benefits to be worth taking the risk. The treatment's effectiveness is something to be determined by the physician, but the quality of life and treatment benefits is a subjective perception and differs from one patient to another.[82] To what degree and how should the information be presented to the patient is not only a matter of science. This presentation involves more profound elements related to the character of the human physician, his clinical acumen, and the virtue of Phronesis.[83]

Before I end up this chapter, I believe it is quite essential to comment on the difference between withdrawing life support treatment, euthanasia, and the notion of intentionality. Sulmasy stated that the value of human life is transcendental, a priori value, and a priori condition for any experience of joy, love, sickness, or suffering. On the other side, the quality of life is a posteriori judgment related to the consequences and perception of life experience. Given that priori state of the value of life, the traditional refrain from killing is a categorical moral obligation in a deontological sense. The intention of euthanasia is to kill the patient, which violates the priori state of the intrinsic transcendental value of life. Intentions precede actions and consequences of the actions. So, if someone accepts the notion that life has intrinsic value, it is quite challenging to justify accepting euthanasia as a way out of a patient's intolerable suffering.[84]

Finally, withdrawing and withholding life-sustaining treatments both fall into a different moral category. The physician's primary intentions here are not to kill the patient but to discontinue a treatment that is now considered a burden on a patient's life. Allowing to die can be morally right or morally wrong, depending on the clinical context.[85] Withdrawing a life-sustaining treatment (extubating a patient) that can save a patient's life and give a meaningful recovery is not permissible and morally wrong. In fact, no one can justify such an action, and I am not sure if there is any intent in such a case other than ending a patient's life. The same withdrawal of life support is morally permissible when there is minimal hope, and the burdens of the treatment outweigh the benefits.

7.6 Summary

This chapter discussed a variety of topics related to the ethics of end-of-life care. Brain death, being one of those controversial subjects, continue to be troublesome to many health care professionals. I recall one of our ICU nurses telling me about a patient she took care of, "She was here for three days on the vent, yet when they withdrew care, they wrote that she died three days ago; that does not make any sense !!!".

This chapter also sheded light on the concept of palliative sedation and the philosophical bases of the doctrine of double effect. It discussed some of the controversies regarding applying this doctrine, especially whether or not we should consider sedation itself an end goal in our care. Lastly, it briefly illustrated the notion of ordinary and extraordinary treatments in the catholic tradition. It emphasized the subjective component in the end-of-life decisions, as each patient's case can have its unique presentation and dimensions, which might change the conduct of our moral algorithm.

Thought Box (VII)
I was quite surprised when I first read and understood the moral reasoning behind palliative sedation. In fact, when I was a resident, our team ordered palliative medications many times on many patients, and such practice did not bring much moral discomfort. Partly because healthcare professionals often tend to follow what others do in training without giving the practice itself much thought. And also, because there was not much focus on educating the staff about the ethical background of palliative sedation. For example, I had no idea about the principle of double effect in all my residency training. Moreover, it was pretty surprising to realize the difference in comfort level between nurses of different institutions regarding administering palliative sedation medications. As if this familiarity with giving these medications and following the protocol is more habitual than intellectual. I am not suggesting that each nurse and physician should become an ethicist to practice end-of-life sedation. However, I believe hospitals should include formal education courses on the moral reasoning of palliative sedation, the differences between killing and allowing to die, and, to less extent, some lecturing on the value of human life and the theories behind the concept. Such knowledge will not only facilitate the workflow but also help offer better care, avoid misconceptions, and clarify the practice from both the staff's and our patients' families' standpoints.

Notes

1. Grof, Stanislav, and Joan Halifax. The human encounter with death. Souvenir Press, 1978. Page 1–5.
2. Sachedina, Abdulaziz Abdulhussein. Islamic Biomedical Ethics: Principles and Application. Oxford: Oxford university press, page 145. Miller AC, Ziad-Miller A, Elamin EM. Brain death and Islam: the interface of religion, culture, history, law, and modern medicine. Chest. 2014;146(4):1092–101. https://doi.org/10.1378/chest.14-0130
3. Jonsen, Albert. "The Birth of Bioethics (New York and Oxford: Oxford University Press)." (1998). Page 235.
4. Jonsen, Albert. "The Birth of Bioethics (New York and Oxford: Oxford University Press)." (1998). Page 236.

5. Silverman, Daniel. "Cerebral Death-The History of the Syndrome and Its Identification." (1971): 1003–1005.
6. onsen, Albert. "The Birth of Bioethics (New York and Oxford: Oxford University Press)." (1998). Page 236.
7. onsen, Albert. "The Birth of Bioethics (New York and Oxford: Oxford University Press)." (1998). Page 237.
8. Nguyen, Doyen. "Pope John Paul II and the neurological standard for the determination of death: A critical analysis of his address to the Transplantation Society." The Linacre Quarterly 84, no. 2 (2017): 155–186. Page 157.
9. Shewmon, D. Alan. "Brain death: can it be resuscitated?." The Hastings Center Report 39, no. 2 (2009): 18–24. Page 18.
10. Beecher, Henry K. "After the definition of irreversible coma." (1969): 1070–1071.
11. onsen, Albert. "The Birth of Bioethics (New York and Oxford: Oxford University Press)." (1998). Page 240.
12. onsen, Albert. "The Birth of Bioethics (New York and Oxford: Oxford University Press)." (1998). Page 243.
13. Korein, Julius. "The problem of brain death: development and history." Ann NY Acad Sci 315 (1978): 19–38. Page 23.
14. Korein, Julius. "The problem of brain death: development and history." Ann NY Acad Sci 315 (1978): 19–38. Page 25.
15. Shewmon, D. Alan. "Mental disconnect:'physiological decapitation'as a heuristic for understanding 'brain death'." The signs of death (2007): 292–333. Page 318.
16. Shewmon DA. "Brainstem death," "brain death" and death: a critical re-evaluation of the purported equivalence. Issues Law Med. 1998;14(2):125–145. Page 127–128.
17. Nguyen, Doyen. "Pope John Paul II and the neurological standard for the determination of death: A critical analysis of his address to the Transplantation Society." The Linacre Quarterly 84, no. 2 (2017): 155–186. Page 167.
18. Kelly, David F., Gerard Magill, and Henk Ten Have. Contemporary Catholic health care ethics. Georgetown University Press, 2013. Page 34–35.
19. Nguyen, Doyen. "Pope John Paul II and the neurological standard for the determination of death: A critical analysis of his address to the Transplantation Society." The Linacre Quarterly 84, no. 2 (2017): 155–186. Page 168.
20. Shewmon, D. Alan. "Mental disconnect: 'physiological decapitation' as a heuristic for understanding 'brain death'." The signs of death (2007): 292–333. Page 320.
21. Di Perri, Carol, Aurore Thibaut, Lizette Heine, Andrea Soddu, Athena Demertzi, and Steven Laureys. "Measuring consciousness in coma and related states." World journal of radiology 6, no. 8 (2014): 589.
22. Bernat JL, Culver CM, Gert B. On the definition and criterion of death. Annals of Internal Medicine. 1981;94(3):389–394, Page 389.
23. Bernat JL, Culver CM, Gert B. On the definition and criterion of death. Annals of Internal Medicine. 1981;94(3):389–394, Page 390.
24. Bernat JL, Culver CM, Gert B. On the definition and criterion of death. Annals of Internal Medicine. 1981;94(3):389–394, Page 389, 391–393.
25. Bernat, James L. "The whole-brain concept of death remains optimum public policy." The Journal of law, medicine & ethics 34, no. 1 (2006): 35–43.
26. Bernat, James L. "A conceptual justification for brain death." Hastings Center Report 48 (2018): S19.
27. Bernat, James L. "A conceptual justification for brain death." Hastings Center Report 48 (2018): S20–S21.
28. Truog, Robert D. "Defining Death: Lessons From the Case of Jahi McMath." Pediatrics 146, no. Supplement 1 (2020): S75–S80. Page 78.
29. Shewmon, D. Alan. "Mental disconnect: 'physiological decapitation' as a heuristic for understanding 'brain death'." The signs of death (2007): 292–333. Page 294.

30. Shewmon, D. Alan. "Mental disconnect: 'physiological decapitation' as a heuristic for understanding 'brain death'." The signs of death (2007): 292–333. Page 295–296.
31. Shewmon, D. Alan. "Brain death: can it be resuscitated?." The Hastings Center Report 39, no. 2 (2009): 18–24. Page 19.
32. Shewmon DA. "Brainstem death," "brain death" and death: a critical re-evaluation of the purported equivalence. Issues Law Med. 1998;14(2):125–145. Page 133.
33. shewmon DA. "Brainstem death," "brain death" and death: a critical re-evaluation of the purported equivalence. Issues Law Med. 1998;14(2):125–145. Page 130–131.
34. Shewmon, D. Alan. "Chronic" brain death": meta-analysis and conceptual consequences." Neurology 51, no. 6 (1998): 1538–1545.
35. Wijdicks, Eelco FM, Panayiotis N. Varelas, Gary S. Gronseth, and David M. Greer. "Evidence-based guideline update: determining brain death in adults: report of the Quality Standards Subcommittee of the American Academy of Neurology." Neurology 74, no. 23 (2010): 1911–1918. Page 1911.
36. Wijdicks, Eelco FM, Panayiotis N. Varelas, Gary S. Gronseth, and David M. Greer. "Evidence-based guideline update: determining brain death in adults: report of the Quality Standards Subcommittee of the American Academy of Neurology." Neurology 74, no. 23 (2010): 1911–1918. Page 1915.
37. Wijdicks, Eelco FM, Panayiotis N. Varelas, Gary S. Gronseth, and David M. Greer. "Evidence-based guideline update: determining brain death in adults: report of the Quality Standards Subcommittee of the American Academy of Neurology." Neurology 74, no. 23 (2010): 1911–1918. Page 1912.
38. Russell, James A., Leon G. Epstein, David M. Greer, Matthew Kirschen, Michael A. Rubin, and Ariane Lewis. "Brain death, the determination of brain death, and member guidance for brain death accommodation requests: AAN position statement." Neurology 92, no. 5 (2019): 228–232.
39. Repertinger, Susan, William P. Fitzgibbons, Mathew F. Omojola, and Roger A. Brumback. "Long survival following bacterial meningitis-associated brain destruction." Journal of Child Neurology 21, no. 7 (2006): 591–595.
40. Truog, Robert D. "Defining Death: Lessons From the Case of Jahi McMath." Pediatrics 146, no. Supplement 1 (2020): S75–S80. Page 77.
41. Shewmon, D. Alan. "Brain Death: A Conclusion in Search of a Justification." Hastings Center Report 48 (2018): S22–S25. Page 23.
42. Wijdicks, Eelco FM, Panayiotis N. Varelas, Gary S. Gronseth, and David M. Greer. "Evidence-based guideline update: determining brain death in adults: report of the Quality Standards Subcommittee of the American Academy of Neurology." Neurology 74, no. 23 (2010): 1911–1918. Page 1916.
43. Shewmon, D. Alan. "Mental disconnect: 'physiological decapitation' as a heuristic for understanding 'brain death'." The signs of death (2007): 292–333. Page 314.
44. Nguyen, Doyen. "Brain death and true patient care." The Linacre Quarterly 83, no. 3 (2016): 258–282. Page 265.
45. Dalle Ave, Anne L., and James L. Bernat. "Inconsistencies between the criterion and tests for brain death." Journal of intensive care medicine 35, no. 8 (2020): 772–780. Page 777.
46. Wijdicks, Eelco FM, Panayiotis N. Varelas, Gary S. Gronseth, and David M. Greer. "Evidence-based guideline update: determining brain death in adults: report of the Quality Standards Subcommittee of the American Academy of Neurology." Neurology 74, no. 23 (2010): 1911–1918. Page 1914.
47. Nguyen, Doyen. "Brain death and true patient care." The Linacre Quarterly 83, no. 3 (2016): 258–282. Page 266–267.
48. Shewmon, D. Alan. "Mental disconnect: 'physiological decapitation' as a heuristic for understanding 'brain death'." The signs of death (2007): 292–333. Page 310–313.
49. Joffe, Ari R., Natalie R. Anton, Jonathan P. Duff, and Allan Decaen. "A survey of American neurologists about brain death: understanding the conceptual basis and diagnostic tests for brain death." Annals of intensive care 2, no. 1 (2012): 4.

50. Pandey, Ashutosh, Pradeep Sahota, Premkumar Nattanmai, and Christopher R. Newey. "Variability in diagnosing brain death at an academic medical center." Neuroscience Journal 2017 (2017).

51. Quill, Timothy E., Bernard Lo, Dan W. Brock, and Alan Meisel. "Last-resort options for palliative sedation." (2009): 421–424. Page 422.

52. Ten Have, Henk, and Jos VM Welie. "Palliative sedation versus euthanasia: an ethical assessment." Journal of pain and symptom management 47, no. 1 (2014): 123–136. Page 124.

53. Kirk, Timothy W., and Margaret M. Mahon. "National Hospice and Palliative Care Organization (NHPCO) position statement and commentary on the use of palliative sedation in imminently dying terminally ill patients." Journal of pain and symptom management 39, no. 5 (2010): 914–923. Page 917.

54. Quill, Timothy E., Bernard Lo, Dan W. Brock, and Alan Meisel. "Last-resort options for palliative sedation." (2009): 421–424. Page 421.

55. Seymour, Jane, Judith Rietjens, Sophie Bruinsma, Luc Deliens, Sigrid Sterckx, Freddy Mortier, Jayne Brown, Nigel Mathers, Agnes van der Heide, and UNBIASED consortium. "Using continuous sedation until death for cancer patients: a qualitative interview study of physicians' and nurses' practice in three European countries." Palliative medicine 29, no. 1 (2015): 48–59.

56. Twycross, Robert. "Reflections on palliative sedation." Palliative Care: Research and Treatment 12 (2019): 1178224218823511. Page 3.

57. Claessens, Patricia, Johan Menten, Paul Schotsmans, and Bert Broeckaert. "Palliative sedation: a review of the research literature." Journal of pain and symptom management 36, no. 3 (2008): 310–333.

58. Inghelbrecht, Els, Johan Bilsen, Freddy Mortier, and Luc Deliens. "Continuous deep sedation until death in Belgium: a survey among nurses." Journal of pain and symptom management 41, no. 5 (2011): 870–879.

59. De Vries, Kay, and Marek Plaskota. "Ethical dilemmas faced by hospice nurses when administering palliative sedation to patients with terminal cancer." Palliative & supportive care 15, no. 2 (2017): 148–157.

60. Putman, Michael S., John D. Yoon, Kenneth A. Rasinski, and Farr A. Curlin. "Intentional sedation to unconsciousness at the end of life: findings from a national physician survey." Journal of pain and symptom management 46, no. 3 (2013): 326–334.

61. Sulmasy, Daniel P. "The use and abuse of the principle of double effect." Clinical Pulmonary Medicine 3, no. 2 (1996): 86–90. Page 87.

62. McIntyre, Alison. "Doctrine of double effect." (2004).

63. Pathan, Hasan, and John Williams. "Basic opioid pharmacology: an update." British journal of pain 6, no. 1 (2012): 11–16. Page 13.

64. Mahler, Donald A., Mahler, Parshall, American Thoracic Society Committee on Dyspnea, O'Donnell, Mahler, von Leupoldt et al. "Opioids for refractory dyspnea." Expert review of respiratory medicine 7, no. 2 (2013): 123–135.

65. Jansen, Kristian, Dagny F. Haugen, Lisa Pont, and Sabine Ruths. "Safety and effectiveness of palliative drug treatment in the last days of life—a systematic literature review." Journal of Pain and Symptom Management 55, no. 2 (2018): 508–521.

66. Sulmasy, Daniel P. "Sedation and care at the end of life." Theoretical medicine and bioethics 39, no. 3 (2018): 171–180. Page 173.

67. Navigante, Alfredo H., Leandro CA Cerchietti, Monica A. Castro, Maribel A. Lutteral, and Maria E. Cabalar. "Midazolam as adjunct therapy to morphine in the alleviation of severe dyspnea perception in patients with advanced cancer." Journal of pain and symptom management 31, no. 1 (2006): 38–47.

68. Sulmasy, Daniel P. "Sedation and care at the end of life." Theoretical medicine and bioethics 39, no. 3 (2018): 171–180. Page 175.

69. Ewen, Alastair, David P. Archer, Naaznin Samanani, and Sheldon H. Roth. "Hyperalgesia during sedation: effects of barbiturates and propofol in the rat." Canadian journal of anaesthesia 42, no. 6 (1995): 532–540. Wang, Qin-Yun, Jun-Li Cao, Yin-Ming Zeng, and Ti-Jun Dai. "GABA ~ A receptor partially mediated propofol-induced hyperalgesia at superspinal

level and analgesia at spinal cord level in rats." Acta Pharmacologica Sinica 25, no. 12 (2004): 1619–1625.

70. Sulmasy, Daniel P. "Sedation and care at the end of life." Theoretical medicine and bioethics 39, no. 3 (2018): 171–180. Page 174.

71. Ten Have, Henk, and Jos VM Welie. "Palliative sedation versus euthanasia: an ethical assessment." Journal of pain and symptom management 47, no. 1 (2014): 123–136. Page 132.

72. Gurschick, Lauren, Deborah K. Mayer, and Laura C. Hanson. "Palliative sedation: an analysis of international guidelines and position statements." American Journal of Hospice and Palliative Medicine® 32, no. 6 (2015): 660–671. Page 668.

73. Quill, Timothy E., and Ira R. Byock. "Responding to intractable terminal suffering: the role of terminal sedation and voluntary refusal of food and fluids." Annals of Internal Medicine 132, no. 5 (2000): 408–414. Page 413.

74. Kirk, Timothy W., and Margaret M. Mahon. "National Hospice and Palliative Care Organization (NHPCO) position statement and commentary on the use of palliative sedation in imminently dying terminally ill patients." Journal of pain and symptom management 39, no. 5 (2010): 914–923. Page 915. The American Academy of hospice and palliative medicine. http://aahpm.org/positions/palliative-sedation.

75. Gurschick, Lauren, Deborah K. Mayer, and Laura C. Hanson. "Palliative sedation: an analysis of international guidelines and position statements." American Journal of Hospice and Palliative Medicine® 32, no. 6 (2015): 660–671. Page 666.

76. Sulmasy, Daniel P. The rebirth of the clinic: An introduction to spirituality in health care. Georgetown University Press, 2006. Page 63.

77. Sulmasy, Daniel P. "Speaking of the value of life." Kennedy Institute of Ethics Journal 21, no. 2 (2011): 181-199. Page 183.

78. Wildes, Kevin W. "Ordinary and extraordinary means and the quality of life." Theological Studies 57, no. 3 (1996): 500–512. Page 500–501.

79. Kelly, David F., Gerard Magill, and Henk Ten Have. Contemporary Catholic health care ethics. Georgetown University Press, 2013. Page 126–129.

80. Wildes, Kevin W. "Ordinary and extraordinary means and the quality of life." Theological Studies 57, no. 3 (1996): 500–512. Page 503–504.

81. Wildes, Kevin W. "Ordinary and extraordinary means and the quality of life." Theological Studies 57, no. 3 (1996): 500–512. Page 507.

82. Pellegrino, Edmund D. "Decisions to withdraw life-sustaining treatment: a moral algorithm." Jama 283, no. 8 (2000): 1065–1067. Page 1066.

83. Pellegrino, Edmund D., and David C. Thomasma. The virtues in medical practice. Oxford University Press, 1993. Page 84–91.

84. Sulmasy, Daniel P. "Speaking of the value of life." Kennedy Institute of Ethics Journal 21, no. 2 (2011): 181–199. Page 184–191.

85. Kelly, David F., Gerard Magill, and Henk Ten Have. Contemporary Catholic health care ethics. Georgetown University Press, 2013. Page 132–134.

Bibliography

1. American Academy of hospice and palliative medicine. http://aahpm.org/positions/palliative-sedation

2. Beecher, Henry K. 1969. After the definition of irreversible coma. 1070–1071.

3. Bernat, James L., Charles M. Culver, and Bernard Gert. 1981. On the definition and criterion of death. Annals of Internal Medicine 94(3): 389–394.

4. Bernat, James L. 2018. A conceptual justification for brain death. Hastings Center Report 48.

5. Bernat, James L. 2006. The whole-brain concept of death remains optimum public policy. The Journal of Law, Medicine & Ethics 34 (1): 35–43.

6. Claessens, Patricia, Johan Menten, Paul Schotsmans, and Bert Broeckaert. 2008. Palliative sedation: A review of the research literature. *Journal of Pain and Symptom Management* 36 (3): 310–333.
7. Dalle Ave, Anne L., and James L. Bernat. 2020. Inconsistencies between the criterion and tests for brain death. *Journal of Intensive Care Medicine* 35(8): 772–780.
8. De Vries, Kay, and Marek Plaskota. 2017. Ethical dilemmas faced by hospice nurses when administering palliative sedation to patients with terminal cancer. *Palliative & Supportive Care* 15(2): 148–157.
9. Di Perri, Carol, Aurore Thibaut, Lizette Heine, Andrea Soddu, Athena Demertzi, and Steven Laureys. 2014. Measuring consciousness in coma and related states. *World Journal of Radiology* 6(8): 589.
10. Ewen, Alastair, David P. Archer, Naaznin Samanani, and Sheldon H. Roth. 1995. Hyperalgesia during sedation: Effects of barbiturates and propofol in the rat. *Canadian Journal of Anaesthesia* 42 (6): 532–540.
11. Grof, Stanislav, and Joan Halifax. 1978. *The human encounter with death.* Souvenir Press, 1–5.
12. Gurschick, Lauren, Deborah K. Mayer, and Laura C. Hanson. 2015. Palliative sedation: An analysis of international guidelines and position statements. *American Journal of Hospice and Palliative Medicine®* 32(6): 660–671.
13. Inghelbrecht, Els, Johan Bilsen, Freddy Mortier, and Luc Deliens. 2011. Continuous deep sedation until death in Belgium: A survey among nurses. *Journal of Pain and Symptom Management* 41 (5): 870–879.
14. Jansen, Kristian, Dagny F. Haugen, Lisa Pont, and Sabine Ruths. 2018. Safety and effectiveness of palliative drug treatment in the last days of life—A systematic literature review. *Journal of Pain and Symptom Management* 55 (2): 508–521.
15. Joffe, Ari R., Natalie R. Anton, Jonathan P. Duff, and Allan Decaen. 2012. A survey of American neurologists about brain death: Understanding the conceptual basis and diagnostic tests for brain death. *Annals of Intensive Care* 2 (1): 4.
16. Jonsen, Albert. 1998. *The birth of bioethics.* New York and Oxford: Oxford University Press.
17. Kelly, David F., Gerard Magill, and Henk Ten Have. 2013. *Contemporary catholic health care ethics.* Georgetown University Press.
18. Kirk, Timothy W., and Margaret M. Mahon. 2010. National Hospice and Palliative Care Organization (NHPCO) position statement and commentary on the use of palliative sedation in imminently dying terminally ill patients. *Journal of Pain and Symptom Management* 39(5): 914–923. Page 915.
19. Korein, Julius. 1978. The problem of brain death: Development and history. *Annals New York Academy of Sciences* 315: 19–38.
20. Mahler, Donald A., Mahler, Parshall, American Thoracic Society Committee on Dyspnea, O'Donnell, Mahler, von Leupoldt et al. 2013. Opioids for refractory dyspnea. *Expert Review of Respiratory Medicine* 7(2): 123–135.
21. McIntyre, Alison. 2004. Doctrine of double effect.
22. Miller, Andrew C., Amna Ziad-Miller, and Elamin M. Elamin. 2014. Brain death and Islam: The interface of religion, culture, history, law, and modern medicine. *Chest* 146(4): 1092–1101. https://doi.org/10.1378/chest.14-0130
23. Navigante, Alfredo H., Leandro CA Cerchietti, Monica A. Castro, Maribel A. Lutteral, and Maria E. Cabalar. 2006. Midazolam as adjunct therapy to morphine in the alleviation of severe dyspnea perception in patients with advanced cancer. *Journal of Pain and Symptom Management* 31(1): 38–47.
24. Nguyen, Doyen. 2016. Brain death and true patient care. *The Linacre Quarterly* 83 (3): 258–282.
25. Nguyen, Doyen. 2017. Pope John Paul II and the neurological standard for the determination of death: A critical analysis of his address to the Transplantation Society. *The Linacre Quarterly* 84 (2): 155–186.
26. Pandey, Ashutosh, Pradeep Sahota, Premkumar Nattanmai, and Christopher R. Newey. 2017. Variability in diagnosing brain death at an academic medical center. *Neuroscience Journal* 2017.

27. Pathan, Hasan, and John Williams. 2012. Basic opioid pharmacology: An update. *British Journal of Pain* 6 (1): 11–16.

28. Pellegrino, Edmund D. 2000. Decisions to withdraw life-sustaining treatment: A moral algorithm. *JAMA* 283(8): 1065–1067. Page 1066.

29. Pellegrino, Edmund D., and David C. Thomasma. 1993. *The virtues in medical practice*. Oxford University Press, 84–91.

30. Putman, Michael S., John D. Yoon, Kenneth A. Rasinski, and Farr A. Curlin. 2013. Intentional sedation to unconsciousness at the end of life: Findings from a national physician survey. *Journal of Pain and Symptom Management* 46 (3): 326–334.

31. Quill, Timothy E., and Ira R. Byock. 2000. Responding to intractable terminal suffering: The role of terminal sedation and voluntary refusal of food and fluids. *Annals of Internal Medicine* 132 (5): 408–414.

32. Quill, Timothy E., Bernard Lo, Dan W. Brock, and Alan Meisel. 2009. Last-resort options for palliative sedation. 421–424. Page 422.

33. Repertinger, Susan, William P. Fitzgibbons, Mathew F. Omojola, and Roger A. Brumback. 2006. Long survival following bacterial meningitis-associated brain destruction. *Journal of Child Neurology* 21 (7): 591–595.

34. Russell, James A., Leon G. Epstein, David M. Greer, Matthew Kirschen, Michael A. Rubin, and Ariane Lewis. 2019. Brain death, the determination of brain death, and member guidance for brain death accommodation requests: AAN position statement. *Neurology* 92 (5): 228–232.

35. Sachedina, Abdulaziz Abdulhussein. 2009. *Islamic biomedical ethics: principles and application*. Oxford: Oxford University Press, Page 145.

36. Seymour, Jane, Judith Rietjens, Sophie Bruinsma, Luc Deliens, Sigrid Sterckx, Freddy Mortier, Jayne Brown, Nigel Mathers, Agnes van der Heide, and UNBIASED consortium. 2015. Using continuous sedation until death for cancer patients: A qualitative interview study of physicians' and nurses' practice in three European countries. *Palliative Medicine* 29(1): 48–59.

37. Shewmon, D.A. 1998. "Brainstem death," "brain death" and death: A critical re-evaluation of the purported equivalence. *Issues in Law and Medicine* 14 (2): 125–145.

38. Shewmon, D. Alan. 2018. Brain death: A conclusion in search of a justification. *Hastings Center Report* 48: S22–S25.

39. Shewmon, D. Alan. 2009. Brain death: can it be resuscitated? *The Hastings Center Report* 39(2): 18–24.

40. Shewmon, D. Alan. 1998. Chronic "brain death": Meta-analysis and conceptual consequences. *Neurology* 51(6): 1538–1545.

41. Shewmon, D. Alan. 2007. Mental disconnect: 'physiological decapitation' as a heuristic for understanding 'brain death'. *The Signs of Death* 292–333.

42. Silverman, Daniel. 1971. Cerebral death-the history of the syndrome and its identification. 1003–1005.

43. Sulmasy, Daniel P. 2018. Sedation and care at the end of life. *Theoretical Medicine and Bioethics* 39 (3): 171–180.

44. Sulmasy, Daniel P. 2011. Speaking of the value of life. *Kennedy Institute of Ethics Journal* 21(2): 181–199. Page 184–191.

45. Sulmasy, Daniel P. 1996. The use and abuse of the principle of double effect. *Clinical Pulmonary Medicine* 3 (2): 86–90.

46. Sulmasy, Daniel P. 2006. *The rebirth of the clinic: An introduction to spirituality in health care*. Georgetown University Press.

47. Ten Have, Henk, and Jos VM Welie. 2014. Palliative sedation versus euthanasia: An ethical assessment. *Journal of Pain and Symptom Management* 47(1): 123–136. Page 124.

48. Truog, Robert D. 2020. Defining death: lessons from the case of Jahi McMath. *Pediatrics* 146 (Supplement 1): S75–S80.

49. Twycross, Robert. 2019. Reflections on palliative sedation. *Palliative Care: Research and Treatment* 12: 1178224218823511.

50. Wang, Qin-Yun, Jun-Li Cao, Yin-Ming Zeng, and Ti-Jun Dai. 2004. GABA~ A receptor partially mediated propofol-induced hyperalgesia at supraspinal level and analgesia at spinal cord level in rats. *Acta Pharmacologica Sinica* 25(12): 1619–1625.

51. Wijdicks, Eelco FM, Panayiotis N. Varelas, Gary S. Gronseth, and David M. Greer. 2010. Evidence-based guideline update: Determining brain death in adults: report of the Quality Standards Subcommittee of the American Academy of Neurology. *Neurology* 74(23): 1911– 1918.
52. Wildes, Kevin W. 1996. Ordinary and extraordinary means and the quality of life. *Theological Studies* 57 (3): 500–512.

Chapter 8
Ethical Dilemmas at the End of Life Addressing Goals of Care

Questions

- Should goals of care be discussed in a paternalistic one-sided model or as a shared decision?
- What are some of the barriers in addressing goals of care?
- How should we respond to a patient family's hope for a miracle?
- How can a physician reliably assess a patient capacity?

8.1 Introduction

Recently, I had a brisk conversation with one of our very smart intensivists in the medical intensive care unit. Not uncommonly, this sort of conversation took place at 2 am after finishing up the second cup of coffee and admitting a long list of very sick patients. After discussing the plan of care for an 88-year-old female who was suffering from respiratory failure and multiple comorbidities, the intensivist asked the resident to call the family regarding goals of care. Of course, he was referring to whether or not they would like us to do CPR and intubate the patient should the clinical situation deteriorated.

The intensivist suggested to the resident to deal with Code Status as a treatment being offered. To act in a mild paternalistic sense and to tell the family that we do not recommend CPR. I was listening to this dynamic teaching session peripherally, attempting to learn something from it. After closing the plan of care, I asked the intensivist whether that was an appropriate approach to discussing goals of care. The intensivist, who was known to be a good physician, and he really is, seemed to be quite confident that this was the best way to approach these conversations. He commented; we studied for many years, we know how the human body works, and we know the odds of success and failure; we recommend and do not recommend therapies all the time, so why should we approach CPR differently?

S. Toro, *Introduction to Clinical Ethics: Perspectives from a Physician Bioethicist*, https://doi.org/10.1007/978-3-031-30804-8_8

The conversation was sharp and direct and did not leave much time for any arguments. Again, it was almost 3 am, we were quite busy, and we had so much work in front of us. After going back to my desk, I paused and thought about the dialogue: Is CPR a treatment in the word's literal meaning similar to other sorts of treatments?

Daniel Sulmasy, a known professor in Bioethics in the United States, suggested a categorization for the different types of therapies that we offer to our patients.[1] He pointed out that all treatments are restorative in nature. They aim at restoring the body's functions. Some of those restorative treatments are regulative such as antipyretics and antiarrhythmic, while the others are constitutive, like dialysis and insulin. Regulative treatments aim to restore the normal physiological states of the body, whereas constitutive therapies aim toward replacing the missing functions of the body. The latter are further divided into replacement (for example, renal transplant) and substitutive (such as dialysis machines and ventilators).

Sulmasy suggested that withdrawing/withholding life support treatments are less controversial if these treatments were regulative and substitutive (1). The problem arises when therapies are considered replacements and become a part of the human body (the long debate about deactivating pacemakers at the end of life). Obviously, CPR does not fit into this category. CPR is not a therapy like Tylenol tablets and insulin injections, uniform, and easy to categorize. In fact, CPR should be considered a package with multiple items (treatments). Some of them could be classified as regulative, like electric shocks and Epinephrine injections. And some are considered substitutive, like Ventilators and ECMO machines. Based on that, our approach to discussing CPR must have some other particularities related to CPR's uniqueness and difference. Additionally, to support my skepticism about the intensivist approach, this debate between the Autonomy-based and Beneficence-based approach in discussing goals of care was studied in a randomized controlled trial at MD Anderson Cancer Center.[2] The study randomized 78 patients into two groups to watch a video module of a conversation between patient-physician discussing code status. The two videos were identical, except that one of them ended with the doctor questioning whether or not the patient would like to get CPR, and the other ended with the physician recommending DNR. The two groups watched the two videos in a different sequence (Q-R, R-Q; R refers to recommendation video, and Q refers to question video). Out of 78 Patients, only 5 changed their opinion regarding DNR. The authors concluded that ending the conversation with an autonomy or beneficence approach did not impact the patient's DNR choice; rather, the conversation's content might be more important in guiding the plan of care.

8.2 Facts and Feelings

8.2.1 General Facts Regarding CPR

The practice of cardio-pulmonary resuscitation started to take place in the 1960s.[3] "Anyone, anytime, can now initiate cardiac resuscitative measures. All that is needed are two hands", these words were written by the father of cardio-pulmonary resuscitation, Professor Leo Kouwenhoven.[4] From a practical standpoint, closed-chest cardiac massage was initially introduced to resuscitate various acute insults, such as electric shocks, drowning, and anesthetic drugs' side effects during operations. Later on, the applications of this technique expanded and became the cornerstone of resuscitating any unfortunate patient who sustained a cardiac arrest (3). Today, CPR courses are widely available to the public, and citizens are encouraged to become certified in basic life support (BLS). Automatic defibrillators are scattered in airports, commercial buildings, gyms, and many other public places.[5] In fact, it has become almost a subconscious norm in society that anyone who sustains a cardiac arrest should receive CPR.

However, the colored picture of CPR is not as quite colorful as it has always been painted. Since the 1970s, physicians started to notice that CPR was not a miraculous technique and the outcomes of CPR were not ubiquitously great. Ethicists also started to raise concerns about the applications of medical technology. In response to that, many physicians initiated the discussion of Do-Not-Resuscitate (DNR) order as a possible option for many sick patients. To complicate the matter more, the patient's right movements added another element to the dilemma of cardiopulmonary resuscitation. These movements increased the focus on patient autonomy and emphasized the right of every person to choose what treatment he wants to receive. Under the influence of these social pressures, the public started to subconsciously perceive DNR as a patient's right to be claimed and not a procedure with specific medical indications.[6]

This brief historical review gives us some idea of why the conversation regarding code status and goals of care became such a complicated issue. In addition to all of that and to the previous ontological analysis of CPR (whether physicians should or should not regard it as a treatment option), I would like to add another point. The public availability and practice of CPR would be a central issue in this debate, even if we agreed on considering CPR a treatment. Take the following analogy. Imagine that Tylenol tablets had a magic effect on reviving patients after cardiopulmonary arrest; in other words, the whole technique of CPR was shrunken down to become a Tylenol tablet. Of course, we are not here discussing the outcomes; I am just referring to the restoration of cardiopulmonary function after CPR. Now let us imagine a physician telling a family we cannot give the patient Tylenol because the likelihood of good outcomes is quite low. Well, the family can go to the market and buy Tylenol tablets themselves. They do not need the doctor's permission to do so. This public availability component is key in this complex conversation. At present, CPR is something that can be offered to any patient by anyone in the community, at least in principle. And many families are not aware of septic shocks, drips, pressors, vents, and other medical issues that complicate CPR in the hospital. What is in their mind is quite different; it is something that can be offered, so why not?

Moreover, studies revealed a poor understanding of cardiopulmonary resuscitation in the general public. Although some CPR components might be known to many, there is a false perception about the prognosis of CPR.[7] For example, in a survey of 50 surrogates of critically ill patients in a tertiary academic center in the USA, only 8% of the participants identified two of the three main CPR components. The majority of them, 72%, thought that survival after CPR was more than 75%.[8] Many other studies confirmed these false high survival expectations. It is well known that TV shows and media have a major effect on the public mind regarding CPR. The survival to hospital discharge rate presented on TV reaches up to 64% compared to 13% cited in scholarly articles.[9] Why is it quite important to know these facts? Because studies have also shown that surrogates tend to change their preferences for CPR once they learned more and understood the effects, burdens, and prognosis of CPR.[10] Especially in the ICU, where the patient population is quite unique in terms of sickness and complexity.

Regarding survival, two studies examined the survival rate post-CPR in ICU patients. The first retrospective cohort looked at the outcomes of 6518 patients who received ICU-CPR. 15% survived to hospital discharge. Only 33% of the survivors were discharged home, less than 5% got discharged home with independent functional status.[11] The second study followed 49,656 patients who sustained CPA in the ICU. The survival rate to hospital discharge was 15.9%. Patients on pressors had worse outcomes compared to those not requiring hemodynamic support. Of those who survived CPR despite being on pressors, only 3.8% were discharged home, and only 3.2% had good neurological outcomes.[12] These findings are consistent with previous surveys, which showed that 1 out of 7 patients who received CPR in the hospital survived to hospital discharge. Of course, the outcomes tend to get worse with comorbidities. For example, in a meta-analysis of CPR outcomes in metastatic cancer patients, none of the 117 patients survived to hospital discharge (9). End state liver disease (ESLD) patients also have a bad prognosis after CPR. In a recent survey of liver patients between 2006 and 2014, the estimated survival to hospital discharge post CPR in ESLD was 10.7%.[13]

Lastly, there are many barriers to effectively addressing goals of care. Chittenden and her colleagues identified some of them. (1) First is lack of time.[14] In health care industry, there is always time pressure on all health care professionals to function efficiently. Time is so precious in light of required productivity. Another issue related to time is the urgency of the conversation. Especially in stressful environments such as the intensive care units, where physicians have to establish rapport and trust with patients in an extremely short period of time. (2) The second barrier is unresolved feelings about death and dying.[15] Modern medicine developed this pseudo absolute power illusion. Medical students and residents continuously learn that the goal of medicine is to cure illnesses and death is the enemy. Of course, no one needs to mention the disasters and the tragedies that resulted from adopting this philosophy. This encounter with the fact of finitude can provoke anxiety in many health care professionals, and many physicians try to avoid mentioning the word death during goals of care discussions. (3) The third barrier is taking away hope. Many health-care professionals fear that telling patients about their bad prognosis might lead to a loss of hope. Whereas the reality showed that most of terminally ill patients know about their poor prognosis and hope for them might not be the cure of their disease, but rather support, less suffering, and reconciliation with family. (4) Finally, lack of training in communication skills and discussing goals of care also play a major role in this matter. In a national survey of 32 accredited U.S residency programs, less than 30% of the program had structured training in the non-pain domain at the end of life, and only a few programs had dedicated assessments and training in the end-of-life domain.[16] Those barriers might explain some of the patient's presenta-tions when someone who is supposed to be DNR-CC end up coming to the ED for syncope during dialysis, or another patient asking for chest compressions without shocks or intubation. Such presentations and many other cases suggest the need for dedicated training and education in end-of-life dilemmas from an ethical, scientific, and practical standpoint.

8.2.2 The Phenomenology of Suffering

Perhaps, there is not a story better than Leo Tolstoy's "The Death of Ivan Ilyich" to refer to while discussing the phenomenology of suffering around the time of death. This epic narrative talks about Ivan, a successful judge in the city of Petersburg in Russia, who was unfortunate to suffer from what seemed to be a cancerous lesion in his abdomen. A sickness state that dramatically changed and altered the course of his life. The story tells us how Ivan started losing weight, getting more fatigued, tired and confused. Day after day, he felt more and more the dread of his vulnerability and finitude.[17] Many significant themes are present in this short powerful narrative. Themes that if we took some time to reflect on, we might realize their presence in our daily encounters with patients. Ivan's disease shifted him toward loneliness.[18] His family and friends abandoned his suffering because it was interfering with their smooth life routine. Congruently, Ivan also isolated himself from the others because he felt that his weakness and illness were now a burden on everyone else, an obstacle to their happiness. The story tells us how Ivan tried to rationally approach his existential finitude through his analogy and thoughts about Caius.[19] Ivan asked if Caius knew how much Vanya (Ivan's nickname) loved the smell of his leather ball and how many times he kissed his mother's hand. How could Caius be mortal in that abstract sense? Ivan knew that all men are mortal, yet he did not think of himself being in place of Caius; Vanya is Mortal did not carry an easy connotation to accept. The idea of mortality was shifted in Ivan's thoughts from being an abstract concept to an instantiation of a character in his final days.

The story also exposed another critical theme in the relationship between Ivan and his doctors. Many of those physicians did not address his real fears. For them, he was a case of floating kidney or a cecum problem.[20] Like many others, he was someone indifferent, a bed number or medical record number in a hospital ward. Ivan was searching for compassion; he looked for someone to understand what he was feeling and what he was worried about. Those themes, which usually shape the life of our very sick patients, are often ignored in the contemporary health care system and moved to the periphery of the care. Not because physicians are careless but because the practice of medicine is no longer focusing on personalized care.[21] Lastly, the story of Ivan tells us how the whole world looked dismal to him when he knew that his disease was serious. As C.S Lewis put it, the experience is no longer in the first mode of consciousness (looking at things) but in the second mode, the mode of living the experience and looking through it.[22] The streets, the people, the buildings, all of them acquired a different meaning in the light of the new lens through which he started to visualize the world.

Svenaeus suggested that the phenomenology of suffering should be understood individually in light of the core life values that a person embodies.[23] For example, the degree of suffering for a pianist after losing one of his fingers might differ from the same injury in a butcher or a soccer player. As Cassel eloquently put it, suffering is individual, and no one can discern why another person is suffering. That is why loneliness is an essential feature of suffering.[24] The narrow contemporary focus on objectivity, and uniformed guidelines had drifted medicine out of its main goal, namely, the relief of suffering. There is no objective way to measure suffering, and the goals of contemporary medicine are not focused on approaching such a mysterious phenomenon.[25] Despite all the pains and tribulations, the conscious existential experience that Ivan endured in his last days was meaningful. He reflected and re-evaluated his grand narrative, his connections with his family and friends, and his relationship with God.[26]

Suffering is part of the mystery of being human. There are no brain centers for suffering or nerve fibers conducting suffering.[27] Suffering is a perceptive experience toward the transcendent. Humans experience this transcendent through three main realms: Intellectual, Moral, and Aesthetic.[28] They realize their limitedness in relationship with those unlimited dimensions. All suffering can be viewed in the relationship between the infinite orientation of humans and the fact of their finitude. Humans search for truths, but they always find themselves being the victims of lies and cheating.[29] In our story, for example, Ivan felt betrayed by his doctors, who lied to him about the seriousness of his disease. Humans are also oriented generally toward knowledge, yet they are confronted with the fact of the infinite nature of knowledge which cannot be totally obtained. Ivan felt his vulnerability when he started losing his sharp mental capacity and started making mistakes. With every decision we make in our life, we choose to lose another possible option; we cannot win it all. That is simply the fact of our existence and relationship with our surrounding reality. Humans are also oriented toward beauty, but they know that their bodies will fail, they will become sick, and all their beauty will fade. Ivan looked at his face and arms in the mirror and noticed the change.[30] He saw the meanings of his brother-in-law's words, "he is a dead man." The brute fact that illnesses and deterioration are part of the reality of life. The fact that death is a destination for everyone. Patients suffer because they sense their mortality, death, and finitude. Finally, humans are also oriented toward a moral realm. They strive to discover, live and act according to this moral code. Yet, they are always confronted with their shortages, flaws, and sins. Ivan reassessed his relationship with his family. In part, this reassessment was not only sociological or emotional but had an objective, right and wrong sort of assessment embodied within it. It is that search for the good and what ought to be done or should have been done, that faces all of us and contributes in part to our pain and suffering. (29)

8.3 Compassion and Trust

8.3.1 The Role of Compassion

If suffering is part of the mystery of our humanity, and it is not totally penetrable to empirical science, then the question remains how do we respond to it? I do not dare to say that I am confident about how to answer this question. However, I believe that Compassion has a profound role in clarifying the perplexity and bewilderment that surrounds suffering.

Compassion refers to the conscious awareness of a suffering human being and a desire to relieve that suffering.[31] The Rousseauian Account of Compassion also emphasizes these ideas, "Compassion is a painful feeling occasioned by the awareness of recognition of someone's suffering or misfortune that triggers action aimed at alleviating the suffering."[32] Etymologically, the word Compassion literally means to suffer together (co-pati).[33] Compassion is a virtue deeply rooted in the Judeo-Christian tradition and especially highlighted in the story of the Prodigal Son and the Good Samaritan.[34] In the first, the youngest son decides to take his inheritance and leaves the family. He recklessly spent all his money and suffered the consequences of his decisions. When he returns, the story tells us, "his father saw him and had compassion, and ran, and fell on his neck and kissed him." The second story, the Good Samaritan, has similar themes. The Samaritan saw a wounded man on the street, being ignored, had Compassion on him, and made a conscious choice to help him.

Compassion is not only a feeling or emotion and should be distinguished from other attributes, namely, Sympathy, Empathy, and Pity.[35] Sympathy is the broadest of the three; it is the human's ability to share feelings with other fellow humans. Sympathy lacks the specificity of compassion, which is directed toward adversity and suffering. For example, someone can sympathize with another person who is quite happy after a great success.

Empathy is the ability of a person to imagine himself in another person's experience. In layman's terms, to put himself in the other person's shoes. Empathy is the first step toward compassion; however, it lacks the "Action-Component" of the latter. An actor can skillfully embody a character she is playing and expresses her emotions and attributes, but not necessarily be compassionate with her. Compassion not only feels but also does. It acts toward the sufferer in an attempt to relieve his suffering.

Finally, there is Pity, which holds a negative connotation, being associated with a sense of condescendence and inequality in the relationship between two humans. Obviously, this attribute has a limited role in the relationship between physicians and patients.

Now from a practical standpoint, the Aristotelian definition of compassion could be quite helpful for many physicians in practice. It asserts two crucial quality themes. First, in order to have compassion for someone the person who is suffering should deserve it (34). Hypothetically, someone can make the case that it is hard to empathize and be compassionate with Jack the Ripper if he was diagnosed with advanced cancer or was beaten hardly by the city police. Although this might sound plausible, taking this Aristotelian condition too seriously can be quite detrimental to many patients. A doctor should be compassionate with a drug addict who is suffering from infective endocarditis even though someone can make the case that the patient is paying the price of his own bad life decisions of using drugs. And similarly, the case can be applied to many other diseases that carry a component of personal responsibility.

The second quality theme in the Aristotelian definition is pivotal. It is the conscious recognition of a person of the possibility of facing the suffering experience himself. It is an imaginative cognitive function and not only an emotion. In the relationship between physicians and patients, Sulmasy refers to this quality when discussing Sir William Osler's famous passage on Aequanimitas. Osler invited physicians to acknowledge their own character's vulnerability and the similarities of patients' foibles and weaknesses with their own. In the same passage, Osler also emphasized the importance of treating patients with compassion and avoiding hardening the human heart by which we all live.[36] The lack of recognition of a physician's own fallibility might lead him to either make dangerous mistakes or projecting his anger and frustrations upon patients.[37]

In the contemporary scientistic environment of medical practice, it is quite common to hear voices advocating for eliminating compassion from the core of health care. In a paper responding to the Francis Report, Anna Smajdor raised many concerns about the role of compassion in delivering health care. The author rejected the intrinsic value of compassion as a virtue in medicine. For her, compassion had to be viewed as a "safety valve," a drive to compel physicians to act to help patients rather than a fundamental theme in the physician–patient relationship.[38] She argues that removing an appendix, emptying a bedpan, or offering food and water to patients does not require much compassion, yet the critical point is completing the task.[39] Moreover, it is quite dangerous to rely mainly on compassion as a drive to deliver care since many physicians are working under a lot of stress and taking care of patients they do not have a close relationship with. How should someone expect physicians to be compassionate with all those people? Smajdor suggested relying on other intelligent management tools, like checklists, to deliver appropriate care rather than focusing mainly on compassion. Furthermore, she raised many concerns about the relationship between compassion and burnout. She argued that no physician could tolerate the emotional burden of having compassion with every patient he sees in the hospital or the clinic. It is emotionally exhausting and totally intolerable.

Though the points mentioned above are compelling, they require close examination of their relationship with the nature of compassion as a virtue.

Edmund Pellegrino discussed two aspects of the virtue of compassion. The first is moral. It is the duty of the physician to be compassionate in the light of the healing relationship. Healing is not only biomedical (removing appendix, giving Augmentin for pneumonia) but also restoring the patient's wholeness. Compassion is necessary for identifying the nuances and seeing the patient's suffering through his own conscious. Without such vision, true healing is not possible. It is true that certain situations might indeed require less compassion than others. However, that is exactly what the core of virtues is all about. Even compassion requires paradoxically a component of Epoche, so a physician does not lose his objective judgment in the light of the emotional stress. Finding the mean between the two far ends of cold disengagement and confusing emotional attachment is the second aspect of compassion, the intellectual one.[40]

A physician might not need to be compassionate in the OR while operating to remove the appendix, but what about before and after surgery? Does compassion play any role? And if compassion is only a safety value, how many safety valves do nurses and physicians have to open every day? Should compassion be only confined to driving health care professionals to act? What about those who are suffering from chronic diseases? Can any physician help them without dwelling on their suffering and altered existential experience?

8.3.2 The Role of Trust

Aside from communicating with empathy and compassion, another attribute plays a significant role in addressing goals of care, namely trust. I frequently grunt humorously when my colleagues ask me during inpatient service transition to address goals of care on a patient that I have never met. In our contemporary, reductionist, mechanistic practice, such a recommendation is not uncommon. Some physicians expect me to call the patient family on the first day of service and tell them to change the code status of their beloved father to DNR!!! Why? because I am a white coated, board certified, physician and I know what I am doing and what is the best for my patients.

Although I might have exaggerated the scenario, many of those who function in our health care system will most likely grasp what I am talking about. To emphasize more on this point, earning patients and families trust is not synonymous with purchasing an item in a store; it is rather a dynamic process that requires continuous presence, effort, and sacrifice.[41] It is true that there is a component of a *Priori* trust, entitled to health care professionals by society and credentialing. However, this component is fragile and can vanish easily if the physician did not respect his distinguished role and his patients' vulnerability. In addition to that, building trust and rapport with families requires time. Especially in a stressful environment such as the ICU, where many of those family members are already passing through unimaginable anxiety. Not taking into consideration the importance of fidelity and the time it requires to be gained can have detrimental effects on health care delivery, especially when addressing code status.

Now let us analyze the concept of trust in more depth. Etymologically, the word *fidelity* has roots in the Latin word *fides*, which involves trustworthiness and veracity.[42] Our trust in each other reduces the complexity of our lives and makes it easier.[43] Imagine if you have to question and distrust everyone around you, to be distrustful to the expiration date of food in the grocery store, distrust the professor who is teaching you, the medications you take, the people who build your apartment, and stability of the chair you are sitting on, what sort of life will you end up living? If someone distrusted everything surrounding him, he would be paralyzed. As John Newman said once, "Life is for action, if we insist on proofs for everything, we shall never come to action, to act you must assume, and that assumption is faith."[44] Faith entails at its core, fidelity, and trust.

From a practical standpoint, the physician–patient relationship has some special characteristics, namely vulnerability and difference in power. It is the trust of the beneficiary (the patient) that the trustee (the physician) will use his power and will for the service of the first and not someone else; that telos of the physician patient relationship which is aiming towards the benefit of the patient and not exploiting him for secondary gains.[45] There is a profound difference in power between the patient and the physician. The patient has to put significant trust in the knowledge and the character of the physician. Trust, in that case, has both intrinsic and instrumental value. The intrinsic value is the respect that will shape the healing relationship as a result of trustworthiness. And instrumental value which is the healing and the benefit that could be successfully delivered to the patient.[46] Many studies showed that trust leads to better outcomes in health care delivery, less anxiety, more patient's autonomy, and more compliance with treatment plans.[47]

Now how do physicians gain patients' trust? Of course, there is no easy answer to this question. I mentioned before the importance of time in building rapport with patients and families. In addition to that, without a doubt, knowledge, credentials, experience, verbal and nonverbal skills are also quite essential. However, there is another component to building trust that I would like to emphasize. It is the idea that the patient is, at the end, a person. The notion of personhood that is often lost in the complexity of contemporary practice, in the abundance of electronic medical records, consults, and problem lists. And especially in the ICU where interaction with patient is often limited by sedation and critical illness.

Now let us elaborate more on the concept of personhood. The traditional western definition of a person is a conscious, self-aware being, able to perceive himself, and able to reflect and interact with the surrounding environment.[48] The word person has roots in the Latin word *persona* and Greek word *proposon*, which refers to a theater mask.[49] A role that a person will play in a plot. This notion of a person entails specificity that distinguishes him or her from another person.[50] Now, how does the situation look in the ICU? Most, if not all, patients are sedated, unconscious, and not interactive. They all seem very similar while intubated. The rooms look very similar, drips are the same, beeping sounds are the same, and the pathologies are very similar. In such an environment, there is a dangerous tendency to dehumanize the depersonalize the experience.[51] That can happen consciously and unconsciously to many health care professionals. Taking that in mind, someone can release how the relationship with a patient's family might become quite problematic. If John became an ARDS case, similar to Cathy, to George, to Mr. Smith to Mrs. White, doctors would unconsciously disengage themselves from the suffering experience of the patient. Such an action will be sensed by the family, and a mutual understanding of the clinical condition would be harder to achieve, which could eventually lead to more suffering and more agony.

One of the main problems with the current approach to discussing goals of care is the tendency to extend this depersonalization phenomenon to the discussion. The recent movement of personalized medicine, which many physicians are quite proud of, is a good example of taking even a more mechanistic reductionist approach in delivering care.[52] When I was a resident, one of my excellent attendings made sure to get as much information as she can about the social life of her patients. She frequently called the families and asked about what their beloved ones used to work, what are their stories, what did they value, how many years they had been married, and many other questions that aimed at humanizing the interaction and relationship with the families. This attempt to engage and dwell in the mystery of the suffering of another person will not only help in gaining trust but will also nourish the experience of the physician himself.[53] Many times, when I discuss code status with families, I disclose to them that one day I will be in their position taking hard decisions on someone I love, a friend, or a family member. It is that acknowledgment of our own vulnerability and mutual experience that open the hearts to accept the coming grief. The acknowledgement that we are at the end, all of us, one human family.

8.4 Difficult Scenarios

8.4.1 Hoping for a Miracle

During my studying years at medical school, I came to know a fellow student who experienced an extraordinary event in his life. This good-hearted gentleman, who later became one of my closest friends, witnessed, along with many of his family members, a miraculous event. One day, the family woke up and noticed that the Virgin Mary's Icon in the house was exuding oil. The news reached out to one of my friend's uncles, who was not quite a believer at that time. This uncle took the icon, removed it from the frame, wiped it, and dried it, attempting to make things logical. Surprisingly, it continued to exude oil and even more than before. My friend told me that his uncle admitted and believed in God that day, and he contributed a large donation to one of the local churches in the city.

I am not writing this short story to convince anyone whether or not miracles do occur. Witnessing any extraordinary event, such as the one mentioned above, is not quite frequent. Yet, there are two pivotal points to emphasize here. The first is that such events are present in the collective memory of our humanity, and they are acknowledged by many despite their scarcity. Second, these events are always perceived by one of our five senses; it is either touched, smelled, tasted, seen, or smelled. That means that despite our encounter with them, someone could always say they might have been an illusion.[54]

Health care professionals, especially those who deal with end-of-life circumstances, frequently encounter families who voice their hopes for a miraculous cure. Before trying to answer the practical aspects of how a physician might deal with these clinical scenarios, I believe it is important to bring into the conversation some theological and philosophical elements.

Many people simply reject the idea of miracles and think it is totally non-sense. The arguments usually flow from two presumptions. The first, is that all those stories, the stories of miracles, were told at times when people were ignorant. Second, after discovering the laws of nature, we know that miracles clearly violate those laws. In his book, The God's Delusion,[55] Richard Dawkins writes, "The nineteenth century is the last time when it was possible for an educated person to admit to believing in miracles like the virgin birth without embarrassment." The response to these claims is not quite difficult. First, many smart contemporary scientists do not agree with Dawkins arguments on miracles, such as Francis Collins, John Polkinghorne, John Houghton, and many others.[56] Second, Dawkins's position is similar to the one of David Hume. It considers miraculous events a violation of the laws of nature. However, our ability to answer such a question does not depend per se on our knowledge of science. The laws of nature tell us how things work under certain conditions. A billiard ball moves in X direction, and reflects in X angle at a certain speed after the application of X driving force. A cube would fall at a certain speed from a tower according to the law of gravitation if no other force affected it. If someone believes that the universe is a close system, and nature cannot be intervened by supernatural power, then the argument of David Hume stands against miracles.[57] But for those who believe that the universe is an open system to God's works and love, miracles can be a manifestation of his grace.[58]

In order to understand this idea more, let us take this famous example from C.S Lewis's article on Science and Religion.[59] Suppose that I put one dollar in my desk drawer today. The next day I came home and put another two dollars in the drawer. According to the laws of arithmetic, I should find a total amount of money of 3 dollars in my drawer that night. Suppose that I woke up the next morning and I found one dollar only, is that a miracle? Had what happened violated the laws of arithmetic? Probably not. It is likely that my brother took two dollars and bought a doughnut or cup of coffee. I recognized the event because I knew the laws of arithmetic, and I can do subtraction. Science can tell us how things work if it was let naturally to happen, it cannot answer whether God can or cannot intervene and change the order of the cause-effect relationship.

Now let us move to some practical facts. Americans, in General, are religious people. According to Harris Poll in 2013, 72% of Americans believed in miracles.[60] In 2007, a survey performed by Pew Forum on Religion and Public involved 35,556 individuals and showed that 79% believed that miracles still occur at present.[61] Another survey done by the same research institution in 2018 showed that 80% of Americans did believe in the presence of a high spiritual being, with 56% of them believing in the God of the bible.[62] Jacobs and Burns surveyed 1006 Americans and 774 trauma physicians regarding the belief in miraculous cures. 61% of the public responders believed that a person with PVS could be saved by a miracle compared to 20.2% of the health care professionals. Moreover, 57.4% of the public responders believed that God could intervene and save a person even when physicians think the patient's case is futile.[63] Another interesting survey, conducted by Zier,[64] interviewed 50 surrogates of crucially ill patients. The survey showed that 18 out of 50 doubted the physician's ability to predict futility based on religious backgrounds and believed that God could intervene and perform a miracle. Boyd[65] also showed that many factors other than physician judgment play a role in determining the families' decisions at the end of life; faith in God was one of them. Furthermore, only 2% of the respondents relied solely on the physicians' prognostic capacity. This data suggests that belief in God's capacity to perform a miracle is not uncommon in clinical practice, especially in the United States. Now, what do people mean when they say we are waiting for a miracle?

Shinall and his colleagues classified the patients' hope for a miracle under four categories: Innocuous, Shaken, Integrated, and Strategic. (1) In Innocuous hope, the hope is plausible but unlikely; no divine intervention is expected. It is a co-existence of denial and acceptance, and middle knowledge, where the patient is aware of the unlikely of a cure, but still denying the certainty of impending death, "The cancer will Melt away!!!!"[66] (2) Shaken hope is the recognition that the miracle is not forthcoming and an expression of anger and frustration toward God. These patients do not reject medical advice. However, they are passing through a spiritual crisis and are having a significant challenge in finding meaning and hope. A support from clergy could be very helpful with spiritual coping in those cases. (3) Integrated: the hope for a miracle is grounded here on a solid religious theology and commitment to a religious community. Citing the scripture might also be present. Sometimes these views put the patient and the family at odds with the team. The medical team should be clear about their professional role in helping and doing no harm when a conflict arises between them and the patient/family. Clarifying the religious community's vision of death and dying and what could be a sign that miracles can help to obtain a better understanding of the patient's values.[67] (4) Finally, Strategic hope: is typically framed with anger and frustration about the care. It is an expression of distrust in the system or treatment team. There is usually a minimal commitment to religious beliefs. It is an attempt to assert power over the care, however, presented in a religious context.[68] The role of the doctor here is to try to find out why the family stopped trusting the team and how this problem can be solved. By taking into consideration the different meanings for hoping for a miracle, a physician might better understand his patient's conditions and family dynamics. Now, is it possible to differentiate authentic belief in miracles from other coping mechanisms?

Although there is no straightforward method to confirm a patient's authentic belief in God, some attributes and behaviors could help in these challenging situations. Sulmasy suggested the following points.[69] (1) Does the patient or the family acknowledge the possibility that the answer to their prayers for a miracle could be No? Religious people do not try to conjure and manipulate God. If the patient is praying and asking God for a miracle but at the same time submitting to God's will, it is likely that his faith is authentic rather than a form of psychological denial. (2) Does the patient/Family acknowledge that God might not need the doctor or a ventilator to perform a miracle? Now, this point is a little bit controversial. One of my colleagues at work once reminded me of that story of the poor man, who kept blaming God for not helping him while continuing to refuse help from all the people around him. When he finally faced God, God told him, I sent all those people in your way to help you, but you refused. Despite the soundness of this point, most, if not all true believers in God, will refrain from forcing medical teams to continue treatments if the they believed that what they are doing is futile. (3) What are the effects of the patient/family beliefs on their all mood, well-being, and relationship with their community? Those in psychological denial or having difficulty trusting the medical teams often express those feelings through disruptive behaviors toward the staff and demand treatments and procedures to be done even when the treatment team thinks it is rather harmful. On the Contrary genuine faith leads the believer to be at peace with God's will. It helps him to cope with his suffering and transcends his painful experience. Lastly, in regard to the relationship with the community, negative coping mechanisms lead to isolation. The more the focus shifts toward continuing aggressive treatments at the end of life, rather than assuring reconciliation with family and beloved ones, the more likely the family is in denial. The more the family is open to considering input from religious authorities and the community of faith, the more likely the matter at hand is a matter of true belief. To emphasize again, true faith does not lead to isolation but rather solidifies the community of believers.

8.4.2   Assessing Capacity

In 1991 the federal government passed the Patient Self Determination Act (PSDA), which stressed the importance of autonomy and advance directive (AD) in patients care.[70] The law stated that all patients admitted to the hospital should be informed about the hospital policy regarding AD, and a copy of this advance directive should be placed in the patient's chart. The purpose of this law was to protect the patient's rights and preferences in case they ended up losing their capacity to make decisions.

For those who practice medicine in the hospital, assessing a patient's capacity can sometimes be quite challenging. Especially in the intensive care unit, where the psychological and biological burdens are at their peak, and many patients are sedated and unable to communicate their wishes and values. For example, studies have shown that patients with Alzheimer's dementia have high rates of incompetence.[71] Not surprisingly, many patients in intensive care units are elderly with some degree of underlying dementia. Depression also affects patients' capacity. Studies have shown that 1 out of 4 patients with depression lack decision-making capacity.[72] Delirium, which is quite common in the ICU, also complicates the assessment of capacity. All those factors, along with family dynamics, substantially affect the decision-making process, and many times, health care professionals find themselves struggling in sorting out the best treatment course and how to proceed in taking care of various sick patients.

Offering an easy, direct, and straightforward way to assess capacity of ICU patients is not practical. The matter at hand is quite complex, as many of us know. However, I do believe that some crucial points are worth discussing here. Appelbaum and Grisso identified four components to assessing capacity.[73] (1) Communicating Choices is one of the necessary but not sufficient conditions for declaring capacity. If a patient cannot communicate his wishes, it would be quite unlikely that he has capacity. Another explanation is that his capacity might have been transiently impaired by medical conditions, for example, sedation, acute psychosis, stroke, or dementia. The physician can ask the following question to assess this point, have you decided which treatment plan you would like to proceed with? Not giving an option at all or moving back and forth between options might be a sign of a lack of capacity. (2) Appreciating the situation is the ability of the patient to acknowledge and understand his current sickness state. When this insight is not present, it is either obscured by some sort of cognitive impairment or psychological denial. To assess this attribute, the physician can ask the following questions: what do you believe is wrong with you? What do you believe would happen to you if you did not get the treatment? (3) The ability to understand is the third component in assessing competence. It is the capacity to grasp the fundamental meaning of the information presented. If a patient could not understand the treatment options, it would be highly unlikely for him to possess capacity to make informed decisions. Also, this component emphasizes that the patient should understand his critical role in the decision-making process. To assess this component, the physician can ask the following questions: Tell me in your words what the doctors told you about your disease. What are the benefits and risks of the treatment option? (4) Manipulating information is the capacity to weigh the burdens and the benefits of the options in the light of the patient's own values. This component is the most complicated and challenging aspect in assessing capacity. It shows how the patient reaches his decision by reasoning and rationality. To assess this function, the physician can ask the following questions: how did you end up choosing this option? Why do you think choice A is better than choice B?[74]

There are another three crucial points related to the concept of capacity that are worth mentioning here. First, capacity should be differentiated from competence which is a legal status determined mainly by the court. When the patient is declared incompetent, a legal guardian is usually assigned to the patient. The latter will be responsible for taking decisions on the patient's behalf. The second point is referred to as the sliding scale. It denotes that patient's capacity should not always be simplified to yes/no answer; however, it should be viewed as a spectrum in the light of the medical conditions and treatment options. For example, a patient might have the capacity to consent for simple low-risk procedures but lack the capacity to make decisions for chemotherapy, angioplasty, anticoagulation, or other high-risk treatments. Also, from a physician's standpoint, the riskier the procedure/treatment is, the more attention is needed to ensure that a patient has capacity to make decisions. Lastly, refusing to accept ordinary treatments should always raise concerns regarding a patient's capacity. Especially when there is a clear indication for the treatment, and the benefits outweigh the possible risks to a great extent. On the other hand, refusal to accept extraordinary treatment is not quite controversial, and it might be grounded on clear wishes and values.[75]

8.5 Summary

This chapter offered a wide approach to discussing goals of care. First, it provided an ontological analysis of the nature of CPR and whether it can be considered a treatment like many other treatments in medicine. Whether or not we ended up considering CPR a treatment, evidence suggests that the paternalistic model, compared to shared decision making, was not superior in end-of-life planning. Moreover, there is a profound role to trust and compassion in addressing goals of care. One of the major problems that we all face these days in our practice is embedded in the depersonalization of the care. It is quite challenging to engage in the experience of our patients and their families if we stayed away from their suffering. Understanding the existential dimension of human suffering will significantly help practicing physicians in approaching these discussions. The more we know about our patients' stories, the more we relate to them, and the more we can gain their trust and offer them help. Lastly, this chapter addressed two common challenging scenarios that frequently face physicians at the end of life, the hope for a miracle and Lack of capacity. Regarding the first, the takeout message is that not all hopes for a miracle are fanatic religious beliefs that cannot be changed. Sometimes these hopes reflect genuine faith in supernatural power. Other times they might be covering psychological denial. Distinguishing between the two conditions is crucial to help our patients and their families. Regarding the second, lack of capacity is also a quite complex notion, especially in critically ill patients. Learning about the four components in assessing capacity, along with the concept of a sliding scale, might also help many practicing physicians in their care. The more critical the treatment or procedure is, in terms of benefits and burdens, the more attention should be given to whether or not a patient has adequate capacity.

Thought Box (VIII)
Before practicing medicine in the United States, I did not come across the term code status. I even barely remember studying the differences between DNR-CC, DNR-CCA, and DNR-Specified while verifying my medical degree and taking board exams. In Syria, there is no such thing as discussing a patient's code status. It is expected that a physician must continue to escalate care to preserve the life of a patient as long as the latter heart is beating. The same is true for palliative sedation. There

isn't such a practice there. In fact, the field itself, is hardly present as a well-defined specialty.

I mentioned this because I want to draw attention to the importance of under- standing how people from different cultures handle goal-of-care discussions. There is often a priori set of expectations from the medical team, and knowing those expecta- tions can help guide the conversation. Also, the faith of the family and the patient can be very influential and understanding how different religions address the preservation of life might help us deliver better care.

Another essential point regarding the same topic: I cannot emphasize more the importance of trust and compassion. In many situations, those are the needed elements when facing challenging clinical scenarios, such as dealing with angry patients and difficult families. If a patient did not trust the doctor, he wouldn't follow his recom- mendations. And this fidelity is applicable to a broad spectrum of situations, from simple ones such as taking a prescribed antiacids to agreeing to receive chemotherapy or major elective surgery. Therefore, before bringing our clinical judgment to the family, we should all try to understand what the family knows, values, and want from us as a medical team.

In addition to all of that, gaining trust is quite challenging with emotional detach- ment, especially when emotions are much present at the bedside. Showing the family that we care about them and their loved ones can change the whole situation. Commu- nicating the understanding of our own vulnerability is essential. I elaborated on this point before that, many times, I tell the family of my patient that one day I will be in their place, making hard decisions on behalf of my loved ones, I feel their pain and understand their fears, I care about their patient, and I am trying my best to help her. This vulnerability is quite powerful. Whether we like it or not, we are all vulnerable, and sharing these emotions unites us because we all know it. We all know that we won't stay forever healthy, we all know that we won't live disease-less lives, and one day all of us will face the end and the final curtain.

Notes

1. Sulmasy, Daniel P. "Within you/without you: biotechnology, ontology, and ethics." *Journal of General Internal Medicine* 23, no. 1 (2008): 69–72.
2. Rhondali, Wadih, Pedro Perez-Cruz, David Hui, Gary B. Chisholm, Shalini Dalal, Walter Baile, Eva Chittenden, and Eduardo Bruera. "Patient–physician communication about code status preferences: a randomized controlled trial." *Cancer* 119, no. 11 (2013): 2067–2073.
3. Blackhall, Leslie J. "Must we always use CPR." *N Engl J Med* 317, no. 20 (1987): 1281–1285. Page 1281.
4. Kouwenhoven, William B., James R. Jude, and G. Guy Knickerbocker. "Closed-chest cardiac massage." *Jama* 173, no. 10 (1960): 1064–1067.
5. Bishop, Jeffrey P., Kyle B. Brothers, Joshua E. Perry, and Ayesha Ahmad. "Reviving the conversation around CPR/DNR." *The American journal of bioethics* 10, no. 1 (2010): 61–67. Page 61.
6. Bishop, Jeffrey P., Kyle B. Brothers, Joshua E. Perry, and Ayesha Ahmad. "Reviving the conversation around CPR/DNR." *The American journal of bioethics* 10, no. 1 (2010): 61–67. Page 62.

7. Heyland, Daren K., Chris Frank, Dianne Groll, Deb Pichora, Peter Dodek, Graeme Rocker, and Amiram Gafni. "Understanding cardiopulmonary resuscitation decision making: perspectives of seriously ill hospitalized patients and family members." *Chest* 130, no. 2 (2006): 419–428.
8. Shif, Yuri, Pratik Doshi, and Khalid F. Almoosa. "What CPR means to surrogate decision makers of ICU patients." *Resuscitation* 90 (2015): 73–78.
9. Chittenden, Eva H., Susannah T. Clark, and Steven Z. Pantilat. "Discussing resuscitation preferences with patients: challenges and rewards." *Journal of hospital medicine* 1, no. 4 (2006): 231–240. Page 235.
10. Murphy, Donald J., David Burrows, Sara Santilli, Anne W. Kemp, Scott Tenner, Barbara Kreling, and Joan Teno. "The influence of the probability of survival on patients' preferences regarding cardiopulmonary resuscitation." *New England Journal of Medicine* 330, no. 8 (1994): 545–549.
11. Gershengorn, Hayley B., Guohua Li, Andrew Kramer, and Hannah Wunsch. "Survival and functional outcomes after cardiopulmonary resuscitation in the intensive care unit." *Journal of critical care* 27, no. 4 (2012): 421–e9.
12. Tian, Jianmin, David A. Kaufman, Stuart Zarich, Paul S. Chan, Philip Ong, Yaw Amoateng-Adjepong, and Constantine A. Manthous. "Outcomes of critically ill patients who received cardiopulmonary resuscitation." *American journal of respiratory and critical care medicine* 182, no. 4 (2010): 501–506.
13. Ufere, Nneka N., Mayur Brahmania, Michael Sey, Anouar Teriaky, Areej El-Jawahri, Keith R. Walley, Leo A. Celi, Raymond T. Chung, and Barret Rush. "Outcomes of in-hospital cardiopulmonary resuscitation for patients with end-stage liver disease." *Liver International* 39, no. 7 (2019): 1256–1262.
14. Chittenden, Eva H., Susannah T. Clark, and Steven Z. Pantilat. "Discussing resuscitation preferences with patients: challenges and rewards." *Journal of hospital medicine* 1, no. 4 (2006): 231–240. Page 234.
15. Chittenden, Eva H., Susannah T. Clark, and Steven Z. Pantilat. "Discussing resuscitation preferences with patients: challenges and rewards." *Journal of hospital medicine* 1, no. 4 (2006): 231–240. Page 232-233.
16. Mullan, Patricia B., David E. Weissman, Bruce Ambuel, and Charles von Gunten. "End-of-life care education in internal medicine residency programs: an interinstitutional study." *Journal of palliative medicine* 5, no. 4 (2002): 487–496.
17. Tolstoy, Leo, graf, 1828–1910. The Death of Ivan Ilych. New York: Bantam Reissue edition, 2004. Page 81.
18. Tolstoy, Leo, graf, 1828–1910. The Death of Ivan Ilych. New York: Bantam Reissue edition, 2004. Page 71.
19. Tolstoy, Leo, graf, 1828–1910. The Death of Ivan Ilych. New York: Bantam Reissue edition, 2004. Page 79–80.
20. Tolstoy, Leo, graf, 1828–1910. The Death of Ivan Ilych. New York: Bantam Reissue edition, 2004. Page 65–66.
21. Cassell, Eric J. "The nature of suffering and the goals of medicine." *Loss, Grief & Care* 8, no. 1–2 (1998): 129–142. Page 260-261.
22. Lewis, Clive Staples. "Meditation in a Toolshed." *God in the dock: Essays on theology and ethics* (1970): 212–15.
23. Svenaeus, Fredrik. "To die well: the phenomenology of suffering and end of life ethics." *Medicine, Health Care and Philosophy* 23, no. 3 (2020): 335–342. Page 336.
24. Cassell, Eric J. "The nature of suffering and the goals of medicine." *Loss, Grief & Care* 8, no. 1–2 (1998): 129–142. Page 275.
25. Cassell, Eric J. "The nature of suffering and the goals of medicine." *Loss, Grief & Care* 8, no. 1–2 (1998): 129–142. Page 276.
26. Micco, Guy, Patrice Villars, and Alexander K. Smith. "The Death of Ivan Ilyich and pain relief at the end of life." *The Lancet* 374, no. 9693 (2009): 872–873. Page 873.
27. Sulmasy, Daniel P. *The healer's calling: A spirituality for physicians and other health care professionals.* Paulist Press, 1997. Page 95.

28. Sulmasy, Daniel P. *The healer's calling: A spirituality for physicians and other health care professionals*. Paulist Press, 1997. Page 97.
29. Sulmasy, Daniel P. *The healer's calling: A spirituality for physicians and other health care professionals*. Paulist Press, 1997. Page 99–101.
30. Tolstoy, Leo, graf, 1828–1910. The Death of Ivan Ilych. New York: Bantam Reissue edition, 2004. Page 74.
31. Zulueta, Paquita de. "Compassion in twenty-first century medicine: Is it sustainable?" *Clinical Ethics* 8, no. 4 (2013): 119–128. Page 119.
32. Ekstrom, Laura W. "Liars, medicine, and compassion." *Journal of Medicine and Philosophy* 37, no. 2 (2012): 159–180. Page 163.
33. Pellegrino, Edmund D., and David C. Thomasma. *The virtues in medical practice*. Oxford University Press, 1993. Page 79.
34. Saunders, John. "Compassion." *Clinical Medicine* 15, no. 2 (2015): 121. Page 121.
35. Pellegrino, Edmund D., and David C. Thomasma. *The virtues in medical practice*. Oxford University Press, 1993. Page 82.
36. Sulmasy, Daniel P. *The healer's calling: A spirituality for physicians and other health care professionals*. Paulist Press, 1997. Page 112–113.
37. Sulmasy, Daniel P. *The healer's calling: A spirituality for physicians and other health care professionals*. Paulist Press, 1997. Page 114.
38. Smajdor, Anna. "Reification and compassion in medicine: A tale of two systems." *Clinical Ethics* 8, no. 4 (2013): 111–118. Page 113.
39. Smajdor, Anna. "Reification and compassion in medicine: A tale of two systems." *Clinical Ethics* 8, no. 4 (2013): 111–118. Page 112.
40. Pellegrino, Edmund D., and David C. Thomasma. *The virtues in medical practice*. Oxford University Press, 1993. Page 80–81.
41. Hendren, Elizabeth M., and Arno K. Kumagai. "A matter of trust." *Academic Medicine* 94, no. 9 (2019): 1270–1272. Page 1270.
42. Corcoran, Blake C., Lea Brandt, David A. Fleming, and Chris N. Gu. "Fidelity to the healing relationship: a medical student's challenge to contemporary bioethics and prescription for medical practice." *Journal of medical ethics* 42, no. 4 (2016): 224-228. Page 227.
43. Pellegrino, Edmund D., and David C. Thomasma. *The virtues in medical practice*. Oxford University Press, 1993. Page 66.
44. Newman, John Henry. *An essay in aid of a grammar of assent*. Catholic publication society, 1870. Page 92.
45. Hui, Edwin C. "The Patient-surgeon Relationship. Part II: Medical Fidelity as Morality and Law." *Asian Journal of Oral and Maxillofacial Surgery* 17, no. 4 (2005): 210-216 (Page 213).
46. McLeod, Carolyn, "Trust", *The Stanford Encyclopedia of Philosophy* (Fall 2020 Edition), Edward N. Zalta (ed.), URL = < https://plato.stanford.edu/archives/fall2020/entries/trust/ > .
47. Dang, Bich N., Robert A. Westbrook, Sarah M. Njue, and Thomas P. Giordano. "Building trust and rapport early in the new doctor-patient relationship: a longitudinal qualitative study." *BMC Medical Education* 17, no. 1 (2017): 1-10. Page 2.
48. Koksvik, Gitte Hanssen. "Silent subjects, loud diseases: enactment of personhood in intensive care." *Health:* 20, no. 2 (2016): 127-142. Page 127.
49. Thomasma, David C., David N. Weisstub, and Christian Hervé, eds. *Personhood and health care*. Vol. 7. Springer Science & Business Media, 2013. Page 13.
50. Thomasma, David C., David N. Weisstub, and Christian Hervé, eds. *Personhood and health care*. Vol. 7. Springer Science & Business Media, 2013. Page 1.
51. Koksvik, Gitte Hanssen. "Silent subjects, loud diseases: enactment of personhood in intensive care." *Health:* 20, no. 2 (2016): 127-142. Page 130.
52. Balducci, Lodovico. "Personhood: an inconvenient truth." (2012): 93–94.
53. Sulmasy, Daniel P. *The healer's calling: A spirituality for physicians and other health care professionals*. Paulist Press, 1997. Page 118.
54. Lewis, C.S. Miracles. Harper Collins, 2009. Page 1–2.
55. Dawkins, Richard. The God Delusion. Black Swan, 2006. Page 187.

56. Lenox, John C. Can Science explain everything? Good Book Company, 2019. Page 76.
57. Lenox, John C. Can Science explain everything? Good Book Company, 2019. Page 80–81.
58. Sulmasy, Daniel P. "What is a miracle?" *Southern medical journal* 100, no. 12 (2007): 1223–1228. Page 1224.
59. Lewis, Clive Staples. *God in the Dock.* Wm. B. Eerdmans Publishing, 2014.
60. Americans' belief in god, miracles and heaven declines. The Harris Poll Web site. http://www.theharrispoll.com/health-and-life/Americans__Belief_in_God__Miracles_and_Heaven_Declines.html. Published December 16, 2013. Accessed September 20, 2016.
61. Pew Research Center. Religion among the mil-lennials: less religiously active than older Americans, but fairly traditional in other ways. 2010. Availablefromhttp://pewforum.org/upl oadedFiles/Topics/Demographics/Age/millennials-report.pdf.
62. Fahmy, Dalia. "Key findings about Americans' belief in God." *Pew Research Center. April* 25 (2018).
63. Jacobs, Lenworth M., Karyl Burns, and Barbara Bennett Jacobs. "Trauma death: views of the public and trauma professionals on death and dying from injuries." *Archives of Surgery* 143, no. 8 (2008): 730–735.
64. Zier, Lucas S., Jeffrey H. Burack, Guy Micco, Anne K. Chipman, James A. Frank, and Douglas B. White. "Surrogate decision makers' responses to physicians' predictions of medical futility." *Chest* 136, no. 1 (2009): 110–117.
65. Boyd, Elizabeth A., Bernard Lo, Leah R. Evans, Grace Malvar, Latifat Apatira, John M. Luce, and Douglas B. White. " "It's not just what the doctor tells me:" factors that influence surrogate decision-makers' perceptions of prognosis." *Critical care medicine* 38, no. 5 (2010): 1270.
66. Shinall Jr, Myrick C., Devan Stahl, and Trevor M. Bibler. "Addressing a Patient's Hope for a Miracle." *Journal of pain and symptom management* 55, no. 2 (2018): 535–539. Page 356.
67. Shinall Jr, Myrick C., Devan Stahl, and Trevor M. Bibler. "Addressing a Patient's Hope for a Miracle." *Journal of pain and symptom management* 55, no. 2 (2018): 535–539. Page 537.
68. Shinall Jr, Myrick C., Devan Stahl, and Trevor M. Bibler. "Addressing a Patient's Hope for a Miracle." *Journal of pain and symptom management* 55, no. 2 (2018): 535–539. Page 538.
69. Sulmasy, Daniel P. "Distinguishing Denial From Authentic Faith in Miracles: A Clinical-Pastoral Approach: Spirituality/Medicine interface project." *Southern medical journal (Birmingham)* 100, no. 12 (2007): 1268–1272.
70. Kelly, David F., Gerard Magill, and Henk Ten Have. *Contemporary Catholic health care ethics.* Georgetown University Press, 2013. Page 180.
71. Appelbaum, Paul S. "Assessment of patients' competence to consent to treatment." *New England Journal of Medicine* 357, no. 18 (2007): 1834–1840. Page 1835.
72. Appelbaum, Paul S. "The MacArthur Treatment Competence Study. III: abilities of patients to consent to psychiatric and medical treatments." *Law and Human Behavior* 19, no. 2 (1995): 149–174. Vollmann, Jochen, Armin Bauer, Heidi Danker-Hopfe, and Hanfried Helmchen. "Competence of mentally ill patients: a comparative empirical study." *Psychological medicine* 33, no. 8 (2003): 1463.
73. Appelbaum, Paul S., and Thomas Grisso. "Assessing patients' capacities to consent to treatment." *New England Journal of Medicine* 319, no. 25 (1988): 1635–1638.
74. Appelbaum, Paul S. "Assessment of patients' competence to consent to treatment." *New England Journal of.Medicine* 357, no. 18 (2007): 1834–1840. Page 1836.
75. Kelly, David F., Gerard Magill, and Henk Ten Have. *Contemporary Catholic health care ethics.* Georgetown University Press, 2013. Page 147.

Bibliography

1. Americans' belief in god, miracles and heaven declines. The Harris Poll Web site. http://www.theharrispoll.com/health-and-life/Americans__Belief_in_God__Miracles_and_Heaven_Dec lines.html. Published 16 Dec 2013. Accessed 20 Sept 2016.

2. Appelbaum, Paul S. 2007. Assessment of patients' competence to consent to treatment. *New England Journal of Medicine* 357 (18): 1834–1840.
3. Appelbaum, Paul S. 1995. The MacArthur treatment competence study. III: Abilities of patients to consent to psychiatric and medical treatments. *Law and Human Behavior* 19 (2): 149–174.
4. Vollmann, Jochen, Armin Bauer, Heidi Danker-Hopfe, and Hanfried Helmchen. 2003. Competence of mentally ill patients: A comparative empirical study. *Psychological Medicine* 33(8): 1463.
5. Appelbaum, Paul S., and Thomas Grisso. 1988. Assessing patients' capacities to consent to treatment. *New England Journal of Medicine* 319 (25): 1635–1638.
6. Balducci, Lodovico. 2012. Personhood: An inconvenient truth. 93–94.
7. Bishop, Jeffrey P., Kyle B. Brothers, Joshua E. Perry, and Ayesha Ahmad. 2010. Reviving the conversation around CPR/DNR. *The American Journal of Bioethics* 10 (1): 61–67.
8. Blackhall, Leslie J. 1987. Must we always use CPR. *The New England Journal of Medicine* 317(20): 1281–1285. Page 1281.
9. Boyd, Elizabeth A., Bernard Lo, Leah R. Evans, Grace Malvar, Latifat Apatira, John M. Luce, and Douglas B. White. 2010. "It's not just what the doctor tells me:" Factors that influence surrogate decision-makers' perceptions of prognosis. *Critical Care Medicine* 38 (5): 1270.
10. Cassell, Eric J. 1998. The nature of suffering and the goals of medicine. *Loss, Grief & Care* 8 (1–2): 129–142.
11. Chittenden, Eva H., Susannah T. Clark, and Steven Z. Pantilat. 2006. Discussing resuscitation preferences with patients: Challenges and rewards. *Journal of Hospital Medicine* 1 (4): 231–240.
12. Corcoran, Blake C., Lea Brandt, David A. Fleming, and Chris N. Gu. 2016. Fidelity to the healing relationship: A medical student's challenge to contemporary bioethics and prescription for medical practice. *Journal of Medical Ethics* 42 (4): 224–228.
13. Dang, Bich N., Robert A. Westbrook, Sarah M. Njue, and Thomas P. Giordano. 2017. Building trust and rapport early in the new doctor-patient relationship: A longitudinal qualitative study. *BMC Medical Education* 17 (1): 1–10.
14. Dawkins, Richard. 2006. *The god delusion.* Black Swan, Page 187.
15. Ekstrom, Laura W. 2012. Liars, medicine, and compassion. *Journal of Medicine and Philosophy* 37 (2): 159–180.
16. Fahmy, Dalia. 2018. Key findings about Americans' belief in God. *Pew Research Center.* April 25.
17. Gershengorn, Hayley B., Guohua Li, Andrew Kramer, and Hannah Wunsch. 2012. Survival and functional outcomes after cardiopulmonary resuscitation in the intensive care unit. *Journal of Critical Care* 27 (4): 421-e9.
18. Hendren, Elizabeth M., and Arno K. Kumagai. 2019. A matter of trust. *Academic Medicine* 94(9): 1270–1272. Page 1270.
19. Heyland, Daren K., Chris Frank, Dianne Groll, Deb Pichora, Peter Dodek, Graeme Rocker, and Amiram Gafni. 2006. Understanding cardiopulmonary resuscitation decision making: Perspectives of seriously ill hospitalized patients and family members. *Chest* 130 (2): 419–428.
20. Hui, Edwin C. 2005. The Patient-surgeon Relationship. Part II: Medical Fidelity as Morality and Law. *Asian Journal of Oral and Maxillofacial Surgery* 17(4): 210–216 (Page 213).
21. Jacobs, Lenworth M., Karyl Burns, and Barbara Bennett Jacobs. 2008. Trauma death: Views of the public and trauma professionals on death and dying from injuries. *Archives of Surgery* 143(8): 730–735.
22. Kelly, David F., Gerard Magill, and Henk Ten Have. 2013. *Contemporary Catholic health care ethics.* Georgetown University Press.
23. Koksvik, Gitte Hanssen. 2016. Silent subjects, loud diseases: enactment of personhood in intensive care. *Health* 20(2): 127–142.
24. Kouwenhoven, William B., James R. Jude, and G. Guy Knickerbocker. 1960. Closed-chest cardiac massage. *JAMA* 173(10): 1064–1067.
25. Lenox, John C. 2019. *Can science explain everything?* Good Book Company.
26. Lewis, C.S. 2009. *Miracles.* Harper Collins.

27. Lewis, Clive Staples. 1970. Meditation in a Toolshed. *God in the dock: Essays on theology and ethics*, 212–215.
28. Lewis, Clive Staples. 2014. *God in the dock*. Wm. B. Eerdmans Publishing.
29. McLeod, Carolyn. 2020. Trust, *The Stanford Encyclopedia of Philosophy* (Fall 2020 edition), Edward N. Zalta (ed.). https://plato.stanford.edu/archives/fall2020/entries/trust/.
30. Micco, Guy, Patrice Villars, and Alexander K. Smith. 2009. The death of Ivan Ilyich and pain relief at the end of life. *The Lancet* 374(9693): 872–873.
31. Mullan, Patricia B., David E. Weissman, Bruce Ambuel, and Charles von Gunten. 2002. End-of-life care education in internal medicine residency programs: An interinstitutional study. *Journal of Palliative Medicine* 5(4): 487–496.
32. Murphy, Donald J., David Burrows, Sara Santilli, Anne W. Kemp, Scott Tenner, Barbara Kreling, and Joan Teno. 1994. The influence of the probability of survival on patients' preferences regarding cardiopulmonary resuscitation. *New England Journal of Medicine* 330 (8): 545–549.
33. Newman, John Henry. 1870. *An essay in aid of a grammar of assent*. Catholic Publication Society.
34. Pellegrino, Edmund D., and David C. 1993. Thomasma. *The virtues in medical practice*. Oxford University Press.
35. Pew Research Center. Religion among the millennials: Less religiously active than older Americans, but fairly traditional in other ways. 2010. Available from http://pewforum.org/uploadedFiles/Topics/Demographics/Age/millennials-report.pdf.
36. hondali, Wadih, Pedro Perez-Cruz, David Hui, Gary B. Chisholm, Shalini Dalal, Walter Baile, Eva Chittenden, and Eduardo Bruera. 2013. Patient–physician communication about code status preferences: A randomized controlled trial. *Cancer* 119(11): 2067–2073.
37. Saunders, John. 2015. Compassion. *Clinical Medicine* 15(2): 121.
38. Shif, Yuri, Pratik Doshi, and Khalid F. Almoosa. 2015. What CPR means to surrogate decision makers of ICU patients. *Resuscitation* 90: 73–78.
39. Shinall Jr, Myrick C., Devan Stahl, and Trevor M. Bibler. 2018. Addressing a patient's hope for a miracle. *Journal of Pain and Symptom Management* 55(2): 535–539.
40. Shinall Jr, Myrick C., Devan Stahl, and Trevor M. Bibler. 2018. Addressing a patient's hope for a miracle. *Journal of Pain and Symptom Management* 55(2): 535–539.
41. Smajdor, Anna. 2013. Reification and compassion in medicine: A tale of two systems. *Clinical Ethics* 8 (4): 111–118.
42. Sulmasy, Daniel P. 2007. Distinguishing denial from authentic faith in miracles: A clinical-pastoral approach: Spirituality/medicine interface project. *Southern Medical Journal (Birmingham)* 100 (12): 1268–1272.
43. Sulmasy, Daniel P. 2007. What is a miracle? *Southern Medical Journal* 100(12): 1223–1228 (Page 1224).
44. Sulmasy, Daniel P. 2008. Within you/without you: Biotechnology, ontology, and ethics. *Journal of General Internal Medicine* 23 (1): 69–72.
45. Sulmasy, Daniel P. 1997. *The healer's calling: A spirituality for physicians and other health care professionals*. Paulist Press.
46. Svenaeus, Fredrik. 2020. To die well: The phenomenology of suffering and end of life ethics. *Medicine, Health Care and Philosophy* 23 (3): 335–342.
47. Thomasma, David C., David N. Weisstub, and Christian Hervé, eds. 2013. *Personhood and health care*. Vol. 7. Springer Science & Business Media.
48. Tian, Jianmin, David A. Kaufman, Stuart Zarich, Paul S. Chan, Philip Ong, Yaw Amoateng-Adjepong, and Constantine A. Manthous. 2010. Outcomes of critically ill patients who received cardiopulmonary resuscitation. *American Journal of Respiratory and Critical Care Medicine* 182 (4): 501–506.
49. Tolstoy, Leo, graf, 1828–1910. 2004. *The Death of Ivan Ilych*. New York: Bantam Reissue edition.
50. Ufere, Nneka N., Mayur Brahmania, Michael Sey, Anouar Teriaky, Areej El-Jawahri, Keith R. Walley, Leo A. Celi, Raymond T. Chung, and Barret Rush. 2019. Outcomes of

in-hospital cardiopulmonary resuscitation for patients with end-stage liver disease. *Liver International* 39(7): 1256–1262.
51. Zier, Lucas S., Jeffrey H. Burack, Guy Micco, Anne K. Chipman, James A. Frank, and Douglas B. White. 2009. Surrogate decision makers' responses to physicians' predictions of medical futility. *Chest* 136 (1): 110–117.
52. Zulueta, Paquita de. 2013. Compassion in 21st century medicine: Is it sustainable? *Clinical Ethics* 8(4): 119–128.

Printed in the United States
by Baker & Taylor Publisher Services

Printed in the United States
by Baker & Taylor Publisher Services